# ECONOMIC REFORMS AND INDUSTRIAL STRUCTURE IN INDIA

# ECONOMIC REFORMS AND INDUSTRIAL STRUCTURE IN INDIA

edited by

SHUJI UCHIKAWA

MANOHAR
2003

First published 2002
First paperback edition 2003



ISBN 81-7304-505-4

*Published by*
Ajay Kumar Jain for
Manohar Publishers & Distributors
4753/23, Ansari Road, Daryaganj
New Delhi 110002

*Typeset at*
Digigrafics
New Delhi 110049

*Printed at*
Lordson Publishers Pvt. Ltd.
Delhi 110007

# Contents

# Preface

Macro economic policies were changed dramatically in the first half of 1990s. At the same time, real GDP grew fast. However, the process of economic reforms stopped in the mid-1990s. At present, second generation economic reforms has become a big issue of economic policies. When economic reforms started in 1991, it was expected that competition would contribute to the improvement of productivity and that resources would be distributed efficiently through market mechanism. On the other hand, there were fears of lopsided development in the consumer durable goods industries catering only to the higher income groups. Although the growth rates of real gross value added in the manufacturing sector were above 10 per cent between 1994-5 and 1995-6, they went down after 1995-6 onwards.

This book tries to examine the effects of economic reforms on used-basis industry groups on the basis of available data. We have investigated the trends of production, investment, employment and labour productivity to identify the reasons for boom and stagnation after the mid-1990s. Each member has his own viewpoint and in some aspects we might have reached different conclusions.

This volume is originally published as *Economic Reforms and Industrial Structure in India* (ASEDP 59) from Institute of Developing Economies-Japan External Trade Organization (IDE-JETRO).

*Mumbai* SHUJI UCHIKAWA
*15 March 2001*

# Introduction

The purpose of this study is to examine the impact of economic reforms on Indian manufacturing sector. Macro economic policies were changed in 1991 after Indian economy faced balance of payment crisis. On 24 July 1991 new industrial policy was announced to promote the private sector by relaxation of regulations. Industrial licensing was abolished in many industries. Besides, areas reserved for public sector were narrowed down and greater participation by private sector was permitted in core and basic industries. The reforms should be seen as being complementary to those undertaken in trade, financial and fiscal policies and in the management of the exchange rate. Trade quantitative regulations have been relaxed since 1991. The exchange rate was unified and rupee rate started to float in 1993. Indian industries were exposed to not only competition in domestic market but also competition with imports. The pressure from WTO encouraged trade liberalization. The wave of globalization swept towards India. 1991 was the turning point of economic policies. However, economic liberalization had started in the 1980s. A comprehensive industrial policy statement was made in July 1980. In 1985 government announced the delicensing of 25 categories of industries. The deregulation was extended to other industries afterwards. Rupee exchange rates had been devalued gradually in the 1980s. On the process, industrial structure changed in some industries. The 1985 textile policy permitted increase of loomage in the mill sector and stopped takeover of sick mills with no reasonable expectation of becoming viable. Regulation of cement industries was removed before 1991. In this study, therefore, we define economic reforms in a broad sense. They include reforms in the 1980s.

Each industry group had different impact. We analyse each used-basis industrial group, i.e. capital goods, intermediate goods, consumer non-durable (CND) goods, and consumer durable (CD) goods industries. They are categorized on the basis of three-digit level industrial classification (Appendix). In capital goods industries, the organized sector is playing important role. Intermediate goods industries are heterogeneous. Economies of scale are important in petrochemical and iron and steel industries. In consumer non-durable goods industries, the unorganized sector is playing an important role in production as well as employment. In consumer durable goods industries, many foreign companies have entered. Authors analyse industries from their points of interests.

R. Nagaraj examines the effects of economic reforms on output, investment, and employment in the first chapter. Total manufacturing sector growth rate during 1992-2000 is slightly lower but statistically significant, compared to that during 1981-91. The slow down after the reforms is mainly in unregistered manufacturing, as the growth rate of the registered sector has remained the same.

Shuji Uchikawa analyses capital goods industries in the second chapter. Investment boom occurred in the first half of 1990s. Some industries in consumer durable and non-durable goods industries grew in the first half of 1990s. Growth of these industries increased demand of intermediate goods through backward linkage. To meet expanding demand, lumpy investment was implemented in the four industry groups between 1994-5 and 1996-7. Large-scale investment created demand for capital goods. Gross value added (GVA) of capital goods industry group accelerated after 1991-2. Production capacity grew faster than demand. Consequently, over capacity made appearance in four industry groups and utilization of capacity declined. Gross fixed capital formation (GFCF) decreased with time lag. Although GVA of capital goods industry group in 1996-7 and 1997-8 was still much higher than GVA in early 1990s, we cannot expect its rapid growth.

Rajendra R. Vaidya investigates the intermediate goods with a special emphasis on cement and steel at both the level of the industry as a whole and at the firm specific level. At the sectoral level we find that in both the industries production, labour productivity and real wages grew rapidly. While capital productivity was stagnant and real prices fell in both the industries. At the firm specific level market shares of leading firms have fallen in both industries. Small firms have considerable difficulty in raising resources to fund investment and this constrains productivity increases in this segment. Financial distress is concentrated in the small industry segment in both cases though this is less pronounced in the steel industry where financial distress is more evenly distributed across size classes. In the case of the steel industry this could lead to pressures to rollback reforms and increase protection in particular and governmental support in general.

Tirthankar Roy examines CND goods industries in the fourth chapter. Given the very nature of the reforms, it is logical to expect that the reforms would affect the CND industries deeply. Most CND industries are labour-intensive, so that export-incentives are expected to benefit them positively, whereas domestic deregulation such as easing protection of small firms is likely to intensify inter-firm competition within CND. Both processes together can be expected, *a priori*, to force some firms to exit and enable the others to improve efficiency. In the organized sector of CND, the post-reform period has seen a distinct improvement in performance and efficiency. On the other hand, the unorganized sector presents a more complex picture. If we confine ourselves to employment, income or output, it is a picture of decline. If we look at labour productivity, unorganized sector has increased in efficiency just as impressively as the organized sector. It is too simplistic to suggest that the unorganized sector as a whole declined in the post-reform period. A more sober conclusion will be that segments of unorganized industry saw shrinking markets, whereas other segments saw adjustments to improve efficiency and competitiveness.

K.V. Ramaswamy analyses CD industries in the fifth chapter. CD industries have grown faster in the post-reform period relative to the pre-reform period. This growth is attributed to the high rates of capital formation due to the liberalization of licensing and entry conditions. Labour productivity, employment and wages are found to have risen faster in a number of CD industries in the reform years. The market structure of CD industries is also shown to have changed with declining concentration in the reform period. The move towards competitive conditions has resulted in the decline of real prices of consumer durables relative to manufactured goods. The impact of economic reforms is welfare enhancing in terms of lower consumer prices and product variety.

It seems that while growth rates of production and labour productivity came up after economic reforms, prices went down. In this context, economic reforms were effective to strengthen competition and improve efficiency. On the other hand, growth rates of production have declined since mid-1990s. Impacts of economic reforms are limited. Economic reforms could not solve over capacity.

SHUJI UCHIKAWA

[illegible] Roy examines CND goods industries in the fourth chapter. Given the new pattern of the reforms, [illegible] expected that the reforms would affect the CND industries [illegible]. Most CND industries are labour-intensive, so that export-incentives are expected to benefit them positively, whereas domestic deregulation such as ending reservation of small firms is likely to intensify inter-firm competition within CND. Both processes together can be expected, [illegible], to make some firms to exit and enable the others to improve efficiency. In the organised sector of CND, the post-reform period has seen a distinct improvement in performance and efficiency. On the other hand, the unorganised sector presents a more complex picture. If we confine ourselves to employment, income or output, it is a picture of decline. If we look at labour productivity, the unorganised sector has increased its efficiency just as impressively as the organised sector. It is [illegible] to suggest that the unorganised sector as a whole declined in the post-reform period. A [illegible] conclusion will be that segments of unorganised industry saw shrinking markets whereas other segments saw adjustment to improve efficiency and competitiveness.

K.V. Ramaswamy examines CND industries in the fifth chapter. CND industries have grown faster in the post-reform period relative to the pre-reform period. This growth is [illegible] (capital [illegible]) due to the [illegible] of licensing, [illegible] labour productivity, employment and wages [illegible] have risen faster than [illegible] of CND industries in the reform [illegible]. The market structure of CND industries is also shown to have changed with [illegible] concentration in the reform period. The [illegible] consumer [illegible] has [illegible] the share of [illegible] consumer durables relative to manufactured goods. The [illegible] competition [illegible] lower prices and product variety.

It appears that while growth rates of production and labour productivity [illegible] the [illegible] economic [illegible] to [illegible] competition [illegible]. On [illegible] rates of [illegible] have [illegible] since 1990s. [illegible] CND [illegible].

# Performance of India's Manufacturing Sector in the 1990s: Some Tentative Findings

R. NAGARAJ

## INTRODUCTION

India's manufacturing sector accounts for about 16 per cent of measured GDP, 10 per cent of total workforce, and a slightly less than 80 per cent of merchandise exports. The registered manufacturing sector constitutes about 2/3rd of value added, and 1/5th of employment. It also represents import competing sector and it is a net importer. The unregistered manufacturing sector contributing 1/3rd of value added and 4/5th of employment, accounts for the bulk of manufactured exports.[1]

The manufacturing sector, though small relative to the economy, even in a comparative perspective, has received an overwhelming attention in India's development effort. In the Mahalanobis strategy—initiated in the mid-1950s that formed the basis for India's planning—development of 'heavy industry' (capital and intermediate goods industries) was vital for maximizing long-term economic growth, and for pursuing economically self reliant, politically autonomous path of development.

Such a structuralist development perspective has been widely criticized in the mainstream economic writings, especially in the face of the exceptional performance of the small, socially homogenous, export-oriented economies of the East and South-East Asia. In the context of the 1991 external payment crisis, the protagonists of the orthodox reforms laid the blame for the crisis to the four decades of import substituting, public sector dominated, industrialization strategy. According to Vijay Joshi and Ian Little, Indian economic strategy, as reflected in various controls, was 'not only microeconomically inefficient but also macroeconomically perverse' (Joshi and Little, 1994 : 3). Therefore, the economic reforms initiated in 1991-2 sought to consciously move away from the earlier development path towards a market-oriented, export-led growth that was implemented as a package of stabilization and structural adjustment programme.

The orthodox reforms initiated since 1991, probably based on the favourable experience with deregulation in the 1980s, dismantled most of the remaining (domestic) industrial output and investment controls, sharply reduced tariffs and removed restrictions on inward foreign direct investment (FDI). In principle, the reforms are expected to accelerate growth in output, investment and manufactured exports, after a brief dip—popularly called the 'J' curve. Further, these policy initiatives, in principle, are expected to improve allocative and productive efficiency.

This paper reports tentative findings on the trends in output and its composition, investment and employment after the reforms, by comparing them with the record of the 1980s. The study uses data from the *National Accounts Statistics* (NAS), *Annual Survey of Industries* (ASI) and the *Net State Domestic Product* (SDP) compiled by the Central Statistical Organisation. Empirical results reported are mainly based on the semi-log trend analysis, and graphical description of the time series. The study will conclude by making a brief and tentative assessment of the performance.

## OUTPUT PERFORMANCE

### *Variation in Yearly Growth Rates*

Figure 1 shows the annual growth rate of manufacturing sector in the 1990s in terms of two indicators, namely the Index of Industrial Production (IIP) for manufacturing and GDP in total manufacturing.[2] Both the series broadly show the same trend: the growth rate dipped sharply in 1991-2 (turning negative in the *NAS* series), and a sharp upturn in the next four years (1992-3 to 1995-6). In fact, the growth rate of 14 per cent in 1995-6 is perhaps the highest that has ever been recorded in the manufacturing sector in India. Such a sharp recovery after the initiation of a programme of stabilization and structural reforms is consistent with the 'J' curve hypothesis.

But, quite contrary to the theoretical prediction, the growth rate fell sharply in the subsequent three years (1996-7 and 1998-9)—

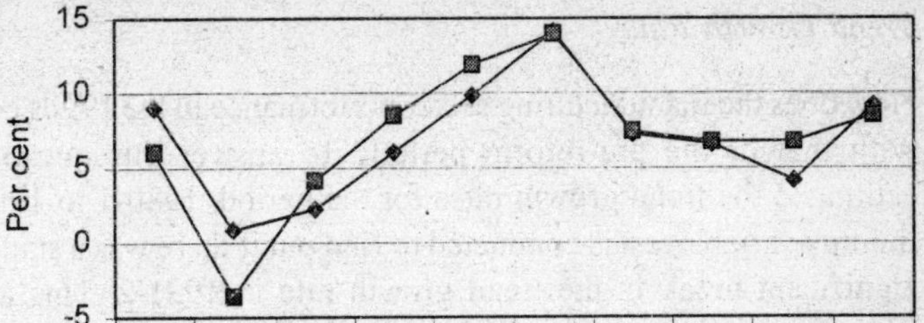


| | 91 | 92 | 93 | 94 | 95 | 96 | 97 | 98 | 99 | 2000 |
|---|---|---|---|---|---|---|---|---|---|---|
| IIP | 8.9 | 0.8 | 2.2 | 6.1 | 9.8 | 14.1 | 7.3 | 6.6 | 4.1 | 9.2 |
| NAS | 6.1 | -3.7 | 4.2 | 8.4 | 11.9 | 14 | 7.4 | 6.7 | 6.7 | 8.5 |

Year ending — IIP — NAS

FIGURE 1. MANUFACTURING GROWTH IN THE 1990s

which is popularly termed as recession.[3] However, the growth rate recovered somewhat in 1999-2000 to about 9 per cent, though its sustainability appears suspect.

TABLE 1. GROWTH RATE OF GDP IN THE TOTAL MANUFACTURING SECTOR, 1980-1 TO 1998-9, IN TWO-DIGIT INDUSTRIAL GROUPS

*(% per year)*

| *Code* | *Industry name* | *1981-99* | *1981-91* | *1992-9* | *Dummy variable* | |
|---|---|---|---|---|---|---|
| | | | | | *Sign* | *Significance* |
| 20-1 | Food | 7.3 | 9.0 | 8.4 | - | Significant |
| 22 | Beverages | 6.0 | 4.8 | 8.7 | + | Not significant |
| 23-6 | Textiles | 2.5 | 4.1 | (-) 1.1 | - | Do |
| 27 | Wood | (-) 0.9 | (-) 2.1 | 3.7 | - | Do |
| 28 | Paper | 7.1 | 8.6 | 6.6 | - | Do |
| 29 | Leather | 6.5 | 4.7 | 6.3 | + | Significant |
| 30 | Chemicals | 10.2 | 9.6 | 10.8 | + | Not significant |
| 31 | Rubber | 11.6 | 15.6 | 8.5 | - | Significant |
| 32 | N.M. minerals | 7.9 | 9.7 | 8.0 | - | Not significant |
| 33 | Metals | 7.6 | 5.2 | 11.6 | + | Do |
| 34 | M. products | 4.3 | 6.2 | 6.9 | - | Significant |
| 35 | N.E. M/c | 5.9 | 6.2 | 7.4 | - | Not significant |
| 36 | Elec. M/c | 9.0 | 12.7 | 7.4 | - | Significant |
| 37 | Transport equip | 7.2 | 6.2 | 12.2 | - | Not significant |
| 38 | Others | 7.3 | 11.4 | 8.7 | - | Significant |
| 39 | Repairs | 5.6 | 6.8 | 8.4 | - | Do |
| | Total | 6.9 | 7.4 | 8.5 | - | Do |

SOURCE: *NAS*, various issues.

*Trend Growth Rates*

How does the manufacturing sector performance in the 1990s compare with that of the pre-reform period? To answer this question, we estimated the trend growth rates for the period, 1980-1 to 1998-9. A dummy variable test is conducted to find out if there was a statistically significant break in the trend growth rate in 1991-2. This analysis was also carried out for GDP in total manufacturing, as well for GDP in registered manufacturing, using the *NAS* data, for 2-digit industry groups (Tables 1 and 2).[4]

Annual growth rates of the total, as well as registered manufacturing GDP during the two decades since 1980-1 are 6.9 and 8.4 respectively. The growth rates for the period before the reforms (1980-1 and 1990-1) are 7.4 per cent and 7.6 per cent per year respectively. The dummy variable test shows there is a statistically significant slow down in the total manufacturing GDP growth rate after 1991-2. However, there are no statistically significant change in the trend growth rate

TABLE 2. GROWTH RATE OF GDP IN THE REGISTERED MANUFACTURING SECTOR, 1980-1 TO 1998-9, IN TWO-DIGIT INDUSTRIAL GROUPS

*(% per year)*

| *Code* | *Industry name* | *1981-99* | *1981-91* | *1992-9* | *Dummy variable* | |
|---|---|---|---|---|---|---|
| | | | | | *Sign* | *Significance* |
| 20-1 | Food | 8.0 | 10.1 | 9.3 | - | Significant |
| 22 | Beverages | 8.2 | 8.6 | 7.6 | - | Not significant |
| 23 | Cotton | 2.5 | 3.6 | 2.8 | - | Do |
| 24 | Wool, silk | 7.6 | 6.2 | 9.9 | + | Do |
| 25 | Jute | 0.6 | (-) 1.8 | 4.1 | + | Do |
| 26 | Textile products | 15.3 | 12.0 | 8.6 | + | Significant |
| 23-6 | Textiles | 5.4 | 4.4 | 6.0 | + | Not significant |
| 27 | Wood | 1.3 | 4.9 | (-) 0.2 | - | Do |
| 28 | Paper | 6.7 | 6.6 | 7.3 | - | Do |
| 29 | Leather | 11.7 | 10.0 | 7.2 | + | Do |
| 30 | Chemicals | 10.4 | 8.9 | 11.7 | + | Do |
| 31 | Rubber | 11.8 | 16.6 | 6.8 | - | Do |
| 32 | N.M. minerals | 8.3 | 10.6 | 4.2 | - | Do |
| 33 | Metals | 7.0 | 4.2 | 15.9 | + | Do |
| 34 | M. products | 5.9 | 4.3 | 12.2 | - | Do |
| 35-6 | N.E. M/c | 7.2 | 7.2 | 8.7 | + | Do |
| 37 | Transport equip | 7.5 | 4.9 | 16.9 | + | Do |
| 38 | Others | 12.7 | 9.3 | 15.9 | + | Do |
| 39+97 | Repairs | 12.5 | 13.9 | 12.2 | + | Do |
| | Total | 8.4 | 7.6 | 11.0 | + | Do |

SOURCE: *NAS*, various issues.

of the registered manufacturing GDP after the reforms. Therefore, the slow down in the total manufacturing growth rate has occurred entirely in the unregistered sector.

In the total manufacturing sector, in terms of two-digit industry groups, leather industry (NIC 29) witnessed an improvement in its growth rate after the reforms. However, the growth rates of other industry groups, viz., food (20-1), rubber (31), metal products (34), electrical machinery (35) and repair services (39), slowed down after 1991-2.

In the registered manufacturing sector, the growth rate of textile products (26) improved after the reforms, while all other industry groups did not experience any change in their growth performance.

Thus, on a trend basis, there is a clear evidence of a modest slow down in the total manufacturing sector growth rate after the reforms. This is entirely in the unregistered sector, as there are no statistically valid changes in the growth rates of the registered sector after the reforms.

*Regional Distribution of Growth*

How is the manufacturing sector growth distributed across the regions? This question is of considerable significance for policy, as it is popularly believed that after the reforms, states that have initiated the market-oriented policies have improved their growth performance. To answer this question, a similar exercise was carried out for the registered manufacturing sector for 17 major states, using the official estimates of net SDP at 1980-1 prices, for the period, 1980-1 to 1997-8, the latest year for which these estimates were available.[5]

The trend analysis shows that there is no statistically significant improvement in the growth rates of these states after 1991-2 (Table 3). But, four states, namely, Bihar, Haryana, Punjab and Uttar Pradesh, experienced statistically significant slow down in their growth rates after the reforms.

An examination of the coefficient of variation (CV) of net SDP in the registered manufacturing sector strengthens this inference (Fig. 2). The CV that was generally declining in the 1980s got reversed in the next decade. In fact, the inequality in 1997-8 is higher than what it was in 1980-1. These trends suggest that although there is no indication of an improvement in the growth rates of any of the major states after the

TABLE 3. GROWTH RATES OF NSDP IN THE REGISTERED MANUFACTURING SECTOR IN SEVENTEEN MAJOR STATES, 1980-1 TO 1998

*(% per year)*

| | *States* | *1981-98* | *1981-91* | *Dummy variable* | |
|---|---|---|---|---|---|
| | | | | *Sign* | *Significance* |
| 1 | Andhra Pradesh | 10.1 | 11.3 | - | Not significant |
| 2 | Assam | 1.8 | 0.7 | - | Do |
| 3 | Bihar | 3.7 | 11.7 | - | Significant |
| 4 | Gujarat | 9.7 | 9.0 | - | Not significant |
| 5 | Haryana | 6.9 | 8.6 | - | Significant |
| 6 | Himachal Pradesh | 15.8 | 19.8 | - | Not significant |
| 7 | Jammu and Kashmir | 7.7 | 7.3 | - | Do |
| 8 | Karnataka | 8.9 | 10.3 | - | Do |
| 9 | Kerala | 4.8 | 4.8 | - | Do |
| 10 | Maharashtra | 7.4 | 7.6 | - | Do |
| 11 | Madhya Pradesh | 8.3 | 8.1 | - | Do |
| 12 | Orissa | 7.2 | 12.9 | - | Do |
| 13 | Punjab | 8.7 | 9.7 | - | Significant |
| 14 | Rajasthan | 9.1 | 9.7 | - | Not significant |
| 15 | Tamil Nadu | 6.2 | 6.4 | - | Do |
| 16 | Uttar Pradesh | 8.5 | 14.2 | - | Significant |
| 17 | West Bengal | 3.2 | 2.4 | - | Not significant |
| | All-India | 7.7 | 8.3 | - | Significant |

SOURCE: *Estimates of Net SDP*, various issues.

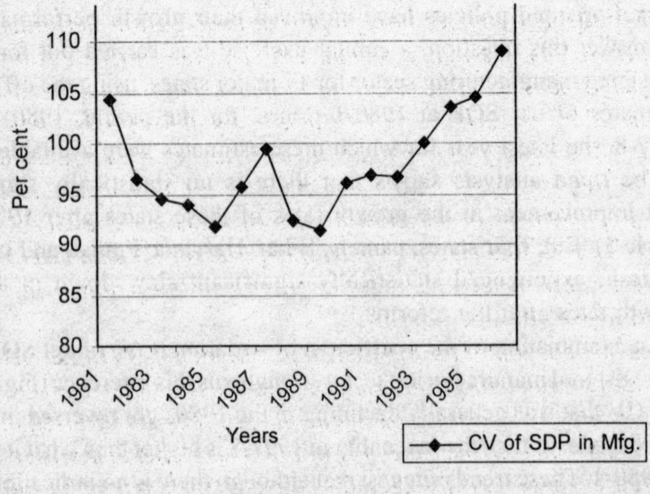


FIGURE 2. REGIONAL DISPARITY IN INDUSTRIAL DEVELOPMENT

reforms, industrial growth seems to be adversely affected in the backward states like Bihar and UP, as well as in the agriculturally advanced states of Punjab and Haryana. Though no evidence is available, one could argue that the observed growing inequality in the manufacturing sector might imply a greater efficiency, as the production decisions are increasingly driven by private profitability considerations.

## INVESTMENT PERFORMANCE

While the growth rate in value added shows no improvement after the reforms—if anything a slow down in the unregistered manufacturing—there has been a boom in gross fixed capital formation (GFCF) or fixed investment, in the registered manufacturing (Fig. 3).[6] The average of the annual growth rate of GFCF went up from about 10 per cent per year in the 1980s to over 15 per cent per year, between 1991-2 and 1996-7. A similar analysis at a disaggregated level suggests that the increase in the rate of fixed investment has occurred across a wide range of industry groups (Table 4). These are textile products (26), wood (27), paper (28), rubber (31), non-metallic minerals (32), metal products (34), transport equipment (37) and repair services (39).

As there has been no perceptible improvement in the economy's aggregate investment rate, the observed change suggests that the composition of fixed investment has shifted in favour of the registered manufacturing sector and away from the rest of the economy—perhaps mainly from infrastructure. Moreover, within the manufacturing sector, GFCF in the unregistered sector has been adversely affected in the 1990s (Fig. 4). In terms of growth rate, it fell from about 7 per cent per year in the 1980s to 5.6 per cent between 1991-2 and 1996-7.

What explains the investment boom when the output growth has not improved? We would hypothesise, the deregulation and the rising business expectations generated by the reform euphoria drove the boom. Perhaps the stock market boom up to the mid-1990s also contributed to it, as it reduced costs of long-term capital.[7]

On the face of it, the decline in fixed investment rate in the unregistered manufacturing sector is puzzling, as the reforms were expected to strengthen the labour-intensive, export-oriented manufacturing, consistent with India's comparative advantage. The reasons for the decline in the fixed investment rate can perhaps be related to

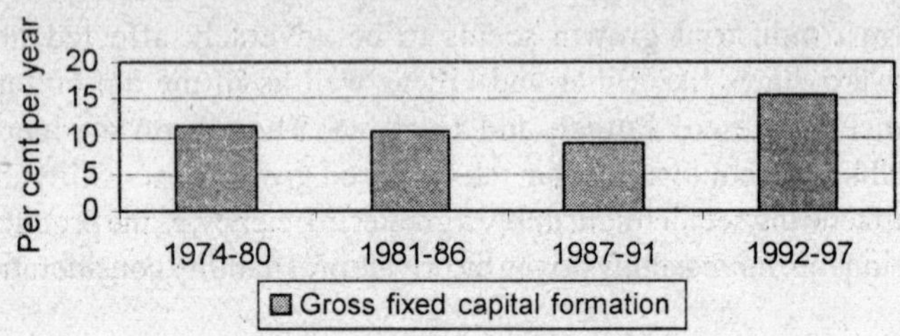


FIGURE 3. FIXED INVESTMENT GROWTH IN REGISTERED MANUFACTURING

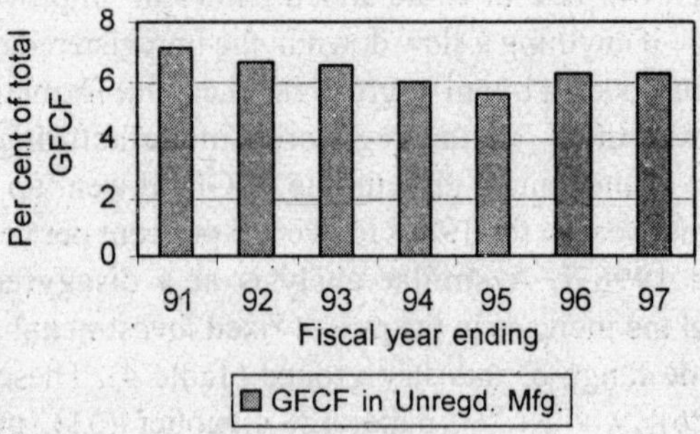


FIGURE 4. UNREGD. MFG.'s SHARE IN TOTAL GFCF

the decline in the credit availability and its high cost in response to the financial sector reforms that was carried out simultaneously. As Figure 5 shows, the share of the small-scale sector in gross bank credit declined from about 13 per cent in 1989-90, to less than 10 per cent in 1996-7.[8]

## EMPLOYMENT PERFORMANCE

While there was much to commend about the improved industrial growth in the 1980s—after nearly one and half decade of slow growth from the mid-1960 to the end of the 1970s—lack of a corresponding employment growth was a major cause for concern. In fact, the registered manufacturing witnessed a decade of 'jobless growth' that attracted considerable analyses, as it was perceived to be central to our understanding of the labour market performance in the organized sector.

The mainstream economic writings suggested that the jobless growth in the 1980s reflected the growing rigidity in the industrial labour market, as employers substituted capital for labour since wages went up sharply (Ahluwalia, 1992; Lucas and Fallon, 1993). However, their views have been widely contested (for instance, Nagaraj, 1994; Bhalotra, 1998). Nagaraj (1994) showed, the rapid rise in workers' earnings in the 1980s represented growth in the mandays worked, while wage rate per manday—for the standard working day—grew at less than one-half the growth in per capita income in this period.

It is in this context that we examine the trends in the registered manufacturing employment in the 1990s. Many serious apprehensions were raised about the potential adverse effects of the reforms on employment. As shown in Figure 6, quite contrary to the trend in the previous decade, there is a steady growth in employment in the 1990s. The reforms in 1991-2 do not seem to have had any influence on this trend. A closer look at the data for the period since 1980-1 suggests, for four years during the mid-1980s (during 1982-3 and 1986-7) employment growth rate was actually negative. Thereafter, there has been a steady growth—at 3.2 per cent per year.

TABLE 4. GROWTH IN FIXED INVESTMENT (GFCF) IN THE REGISTERED MANUFACTURING SECTOR IN TWO-DIGIT INDUSTRIAL GROUPS, 1981-98

*(% per year)*

| *Code* | *Industry group* | *Average of annual growth rates* | | |
|---|---|---|---|---|
| | | *1982-91* | *1982-98* | *1992-8* |
| 20-1 | Food | 13.2 | 10.5 | 6.6 |
| 22 | Beverages | 21.9 | 16.7 | 9.4 |
| 23 | Cotton textiles | 4.7 | 8.7 | 14.3 |
| 24 | Wool, silk textiles | 15.2 | 14.8 | 14.3 |
| 25 | Jute textiles | n.a | n.a | 9.0 |
| 26 | Textile products | 17.3 | 21.2 | 26.7 |
| 23-6 | Textiles | 6.5 | 9.6 | 14.0 |
| 27 | Wood | 35.7 | 28.5 | 18.2 |
| 28 | Paper | 10.0 | 22.1 | 39.3 |
| 29 | Leather | 18.6 | 12.6 | 3.9 |
| 30 | Chemicals | 10.4 | 14.9 | 21.3 |
| 31 | Rubber | 19.5 | 21.2 | 23.6 |
| 32 | N.M. minerals | 11.0 | 13.1 | 16.1 |
| 33 | Metals | 11.8 | 10.0 | 7.4 |
| 34 | Metal products | 8.7 | 12.6 | 18.3 |
| 35-6 | N.E. M/c | 10.7 | 8.7 | 5.9 |
| 37 | Transport equipment | 7.6 | 13.8 | 22.6 |
| 38 | Others | 18.0 | 21.2 | 25.7 |
| 39+97 | Repair services | 13.9 | 35.2 | 22.7 |
| | Total | 17.3 | 12.3 | 15.3 |

SOURCE: *ASI*, various issues.

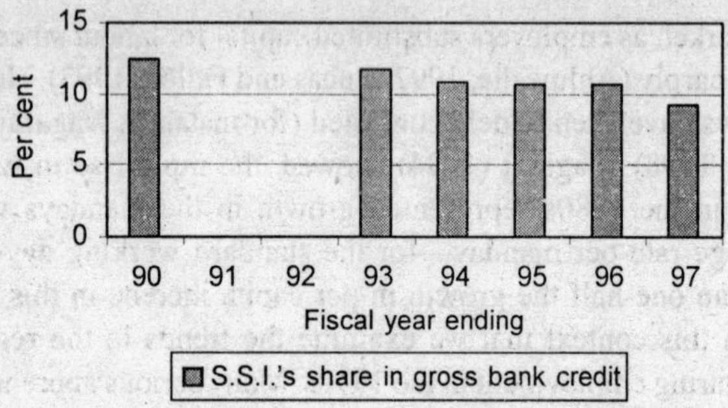


FIGURE 5. S.S.I.'s SHARE IN GROSS BANK CREDIT

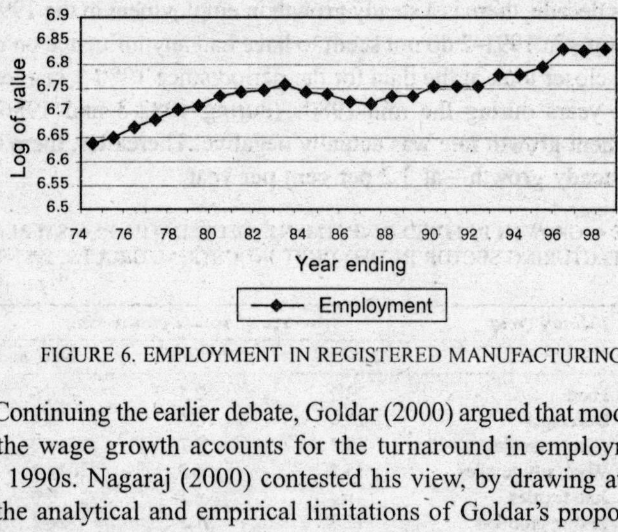


FIGURE 6. EMPLOYMENT IN REGISTERED MANUFACTURING

Continuing the earlier debate, Goldar (2000) argued that moderation in the wage growth accounts for the turnaround in employment in the 1990s. Nagaraj (2000) contested his view, by drawing attention to the analytical and empirical limitations of Goldar's proposition.

While employment trends in the registered sector attracted so much attention, perhaps more significant changes were taking place in the unregistered manufacturing that went unnoticed. An analysis of economic census data for the period 1978-9 to 1994-5 suggests a negative growth rate in employment as well as in the number of establishments between 1984-5 and 1994-5 (Table 5).* This observation seems broadly consistent with the slow down in growth in output

* Admittedly this finding of a fall in employment is at variance with the NSS employment-unemployment survey data. Therefore, the data may be suspect. But the economic census data on establishment where quality is not in question, also shows a corresponding fall. For a careful analysis of the trends in the 1990s, it is necessary to contend with these divergent trends.

TABLE 5. GROWTH IN NUMBER OF ESTABLISHMENTS AND EMPLOYMENT IN UNREGISTERED MANUFACTURING, 1985 TO 1994-5 FOR ALL OVER INDIA

*(percentage change over the previous period)*

| *Year* | *Establishments* | *Employment* |
|---|---|---|
| 1978-9 | | |
| 1984-5 | 15.1 | 12.8 |
| 1989-90 | 3.9 | -0.8 |
| 1994-5 | -9.5 | -1.3 |

SOURCE: Lalitha (1999).

and fixed investment in unregistered manufacturing that was earlier discussed.

There is probably more to the fall in the employment growth rate in the unregistered manufacturing sector than the slow down in the output and investment growth. Perhaps, the fall in employment growth was also because of the real incidence of adjustment problems to reforms faced in the organized sector, being passed on to workers and enterprises in the unorganized sector as the sub-contracting relationship between the two sectors had grown during the last two decades.[9]

In sum, the manufacturing employment scenario after the reforms is perhaps not bright, although the registered manufacturing employment turned around. There are competing explanations for the turnaround, requiring further research. Perhaps, the real brunt of the adjustment is borne by the workers in the unregistered sector which witnessed negative growth in employment and in the number of enterprises between the mid-1980s and the mid-1990s.

## SUMMARY AND CONCLUSION

This preliminary study on India's manufacturing sector since the reforms of 1991-2 examined three aspects of performance, namely, output (value added), fixed investment and employment growth. This assessment is broadly based on comparing the performance after the reforms with that of the 1980s. The analysis is mainly based on NAS, ASI, and the SDP data, disaggregated by two-digit industry groups, as well as, by the registered and unregistered manufacturing.

In the literature on the economic reforms, it is widely believed that the industrial growth follows a 'J' curve—that is, output growth contracts immediately after initiating the stabilization and structural adjustment programme. However, as the structural reforms take root, output growth rebound rapidly as supply responds to market signals in the deregulated environment. We examine if such a process of adjustment was evident in India in the 1990s.

As predicted, output growth dipped immediately after the reforms, but it turned around quickly to follow a 'J' curve during the four years since 1992-3. The growth rate in 1995-6, at 14 per cent, was the highest that was ever recorded. However, such a growth was not sustained, as there was a setback in the second half of the 1990s.

On a trend basis, therefore, the total manufacturing sector growth rate during 1992-9 is slightly lower but statistically significant, compared to that during 1981-91. The slow down after the reforms is entirely in the unregistered manufacturing, as the growth rate of the registered sector has remained the same.

The only industrial groups that experienced an improvement in the growth rate after the reforms are leather (NIC 29) in the total manufacturing sector, and textile products (26) in the registered manufacturing sector. However, a much wider range of industry groups, both in the total and the registered manufacturing sectors, witnessed a slow down in their growth rates after the reforms.

How did the reforms affect industrial growth across the states? This is an important question for policy, as it is widely believed that the states that have initiated the reforms improved their performance. Our analysis suggests that there is no evidence of a statistically significant improvement in the registered manufacturing growth rates in any of the seventeen major states of India after the reforms. But the growth rates in four industrially backward and agriculturally advanced states of Bihar, UP, Punjab and Haryana actually witnessed a slow down after the reforms. Therefore, inequality in industrial output across the states widened in the 1990s, reversing the declining trend of the previous decade.

Thus, on output performance, it seems reasonable to infer that the manufacturing growth rate evidenced a slow down after the reforms. This slow down affected the unregistered sector, a sizeable cross section of the industrial groups, and four states in north India. Inter-regional inequality in industrial growth turned adverse in the 1990s, after a decade of equalising growth. Whether the pattern of growth has improved efficiency is a moot point.

There has been a boom in fixed investment in the 1990s. But it is entirely in the registered sector, though widely distributed across the industry groups. However, the growth of investment in labour-intensive, export-oriented, unregistered manufacturing sector came down after the reforms. Abolition of various output and investment controls, and rising business expectations generated by the reforms seem to be responsible for the growth in investment. This was perhaps facilitated by the stock market boom in the first half of the 1990s that reduced the cost of long-term capital.

By most popular indicators, the reforms increased domestic and import competition with a favourable impact on relative prices. This was perhaps best illustrated by relative decline in capital goods prices for the first time in the 1990s (Nagaraj, 1997).

Investment boom in the registered manufacturing turned around employment growth, after a decade of stagnation. But, the growth in employment and in number of enterprises suffered in the unregistered sector, perhaps bearing the real brunt of adjustment in the organized sector—considering the close inter-firm linkages between the organized and the unorganized sectors in the product and labour markets. Perhaps, this also reflects the decline in credit availability and increase in its cost for small enterprises in response to the financial sector reforms that were simultaneously initiated in the 1990s.

Have the reforms enabled the manufacturing sector to overcome the microeconomic inefficiency and the macroeconomic perversity that Vijay Joshi and Ian Little talked about? Probably not, for the following reasons:

- The flip side of the investment boom without a corresponding increase in output growth was the increase in the levels of excess capacities and increase in incremental capital output ratio.
- As there was no improvement in machanise export growth on a trend basis in the 1990s, macroeconomic vulnerability persisted.
- Much of the fixed investment growth accrued to the import competing sector, rather than for building export capability. Therefore, the economy's vulnerability to external shocks was not reduced.

In sum, the reforms surely brought in some notable improvements in the functioning of industrial markets due to greater competition and quality improvements. But, the reforms neither improved the output nor the employment growth rates, in the total manufacturing sector. The distribution of the growth across regions and between the organized and the unorganized sectors turned adverse. And above all, it seems hard to believe that the reforms enabled India to overcome the 'macroeconomic perversity', as India's exports or export capability have not improved on a sustained basis, at least not as yet.

## NOTES

1. Registered manufacturing mostly consists of all factories employing 10 or more workers (20 or more workers without using power) using power. Unregistered manufacturing refers to those manufacturing activities that are outside the registered sector.
2. There has been a change in the base year of IIP and of *NAS*, to 1993-4, replacing the earlier series with 1980-1 as the base year. We have merged the two series to create consistent time series for the 1990s. *NAS* estimates for 1999-2000 are quick estimates, and may undergo revision. IIP has serious limitations as its quality has deteriorated over the decades (Nagaraj, 1999). Yet, the data has some value at one digit level, as they are available with least time lag.
3. Strictly speaking, recession is defined as two consecutive quarters of negative growth rate. India did not witness negative growth rate in any quarter.
4. Unless otherwise stated, all variables in this study are at constant prices.
5. We did not use figures for the total manufacturing, as the state level unregistered manufacturing value added estimates reportedly have many problem of quality.
6. Shuji Uchikawa's paper in this project provides detailed quantitative picture of this boom.
7. Although there is no statistically valid association between GFCF and stock market mobilization in the long run (Nagaraj, 1996), in the short to medium term, the relationship could be different.
8. It is widely known that small entrepreneurs are severely credit constrained; fixed investment in small enterprises in developing economies is largely a function of timely availability of credit.
9. This is perhaps similar to what was observed in the Japanese economy, in the 1970s, where the real brunt of adjustment to market fluctuation were borne by workers in the small firms (Dore, 1986).

## REFERENCES

Ahluwalia, Isher Judge, *Productivity and Growth in Indian Manufacturing*, New Delhi: Oxford University Press, 1992.

Bhalotra, Sonia R., 'Puzzle of Jobless Growth in Indian Manufacturing', *Oxford Bulletin of Economics and Statistics*, Vol. 60, No.1, January 1998.

Dore, Ronald, *Flexible Rigidities: Industrial Policies and Structural Adjustment in the Japanese Economy, 1970-80*, London: Althone Press, 1986.

Fallon, Peter and Robert E.B Lucas, 'Job Security Regulation and the Dynamic Demand for Industrial Labour in India and Zimbabwe', *Journal of Development Economics*, Vol. 40, No. 2, April-June 1993.

Goldar, Bishwanath, 'Employment Growth in Organised Manufacturing in India', *Economic and Political Weekly*, Vol. 35, No. 14, 1-7 April 2000.

Joshi and Little, *India: Macroeconomics and Political Economy, 1961-1991*, New Delhi: Oxford University Press, 1994.

Lalitha, N., *Unorganised Manufacturing Sector in the National Employment and Gujarat State*, report submitted to NCAER, New Delhi, 1999.

Nagaraj, R., 'Organised Manufacturing Employment', *Economic and Political Weekly*, Vol. 35, No. 38, 16 September 2000.

_______ 'How Good are India's Industrial Statistics? An Exploratory Note', *Economic and Political Weekly*, Vol. 34, No. 6, 6-12 February 1999.

_______ 'What has happened since 1991? Assessment of India's Economic Reforms', *Economic and Political Weekly*, Vol. 32, Nos. 44-5, 8-14 November 1997.

_______ 'India's Capital Market Growth: Trends, Explanations, and Evidence', *Economic and Political Weekly*, Vol. 31, Nos. 35-7, September 1996.

_______ 'Employment and Wages in Manufacturing Industries: Trends, Hypotheses and Evidence', *Economic and Political Weekly*, Vol. 29, No. 4, 22 January 1994.

# Investment Boom and Capital Goods Industry

SHUJI UCHIKAWA

The recent economic reforms have affected the manufacturing sector on four counts. First, with the abolition of industrial licensing, private sectors were able to build and expand their capacity without any regulation. Production of consumer durable goods started to meet the pent-up demand, which was not available under import-substitution policies.[1] Second, import quantitative regulation was relaxed and Indian manufacturers could therefore import capital and intermediate goods easily. Third, growth rate of public investment has decelerated. During the 1980s, public investment played an important role to generate economic growth. And with the start of economic reforms in 1991, the share of the private corporate sector in total investment increased. Fourth, private sector now had the option to procure money from financial market in addition to bank lending. This chapter examines the kind of impact economic reforms had on the capital goods industry.

## 1. INVESTMENT BOOM AFTER ECONOMIC REFORMS

During the first half of the 1980s, the central government allotted relative high percentage of the budget to public investment. Shares of capital outlay in total expenditure of the central government were around 15 per cent. However, accumulated budget deficits made the central government issue a large number of government bonds which became a burden afterwards. After 1987-8 interest payment exceeded capital outlay. The share of capital outlay went down to below 8 per cent after 1995-6. The private corporate sector increased investment rapidly and as a result, trends of gross fixed capital formation (GFCF) have accelerated in the 1990s.

Since Annual Survey of Industries (ASI) does not have used-based classification, we must aggregate three digit level data to produce used-based data series. National industrial classification (NIC) was changed from 1970 code to 1987 code. After it was changed, ASI changed the classification in 1989-90 to follow the new classification. This study produces used-based data series on the basis of 1987 code. Gross value added (GVA) is deflated by chain index of industry group. Annual inflation rate of specific year, denoted by $P_t$, is obtained using the formula:

$$P_t = \sum \frac{WPI_{i,t} V_i}{WPI_{i,t-1} V_a}$$

where $V_i$ is net value added (NVA) of good i and $V_a$ is total NVA of the industry group. Wholesale price index of good i is compared with previous year's and is weighted by the share of NVA in total NVA of the industry group. Index takes 1981-2 as the base year. GFCF is deflated by chain index of capital goods. GFCF at the 1981-2 prices accelerated after 1991-2 in all four industry groups (Fig. 1). Annual average of GFCF of total industries at 1981-2 prices rose from Rs 7,191 crore between 1986-7 and 1990-1 to Rs 14,438 crore between 1991-2 and 1995-6. As mentioned later, the investment boom was accompanied by increase of GVA of capital goods industries. Thus only a boom of consumer durable goods cannot explain the development of capital goods industries.

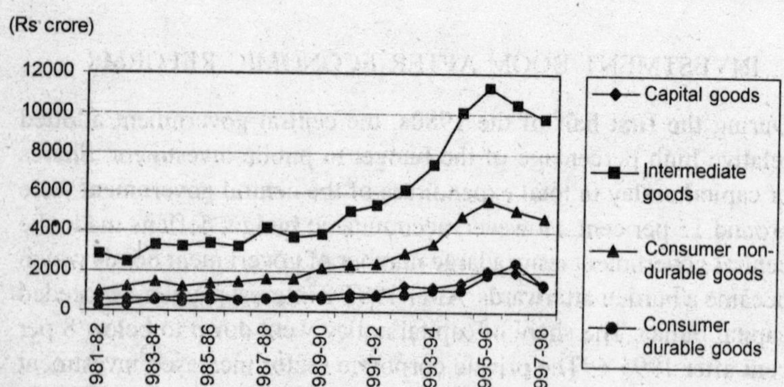


FIGURE 1. GFCF IN 1981-2 PRICES

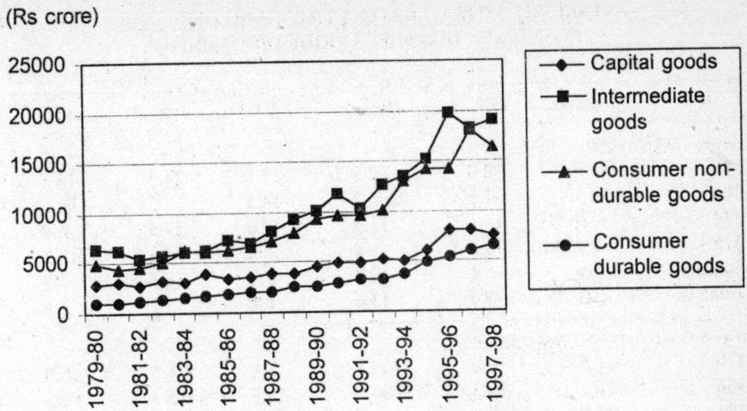


FIGURE 2. GVA IN 1981-2 PRICES

One can estimate four reasons for the investment boom. First, if market shows sign of good prospects, capacity will be expanded by the expectation. Second, financial liberalization made it easy for private corporate to procure fund. Third, if internal financing increases, it might encourage investment. Fourth, if relative price of capital goods vis-à-vis product price decreases, firms might increase investment.

GVA is a good indicator of market conditions. As rapid growth of GVA in an industry generates expectation that the industry will grow in the future, investment in the industry might increase. Abolition of industrial licensing might have encouraged investment on the basis of entrepreneurship. But entrepreneurs have their own estimation of market prospects and it does not always depend on the present trends of production, though some of them might have failed due to a wrong judgement. However, we can safely say that the main stream of investment will result in the rise of industries, not its decline.

Growth rates of GVA accelerated after 1991-2 (Fig. 2). All four industry groups and total industries pass the dummy test with significant at 1 per cent.[3] Investment increased to fill up the expanding demand. In fact, growth rates of GFCF of all four industry groups and total industries accelerated between 1991-2 and 1995-6. We analyse which industries are growing in each industry group. When we consider backward effects, condition of consumer goods is important.

TABLE 1. GROWTH RATES OF GVA AND GFCF IN CONSUMER DURABLE GOODS INDUSTRIES

(%)

| *Code* | *1992-3* | *1993-4* | *1994-5* | *1995-6* | *1996-7* | *1997-8* |
|---|---|---|---|---|---|---|
| GROSS VALUE ADDED | | | | | | |
| 310 | 28.4 | 0.0 | 12.8 | -13.1 | 31.4 | 14.2 |
| 366 | 16.5 | -7.3 | 200.7 | -44.8 | 2.2 | 63.3 |
| 374 | -6.9 | 45.1 | 11.4 | 47.0 | 27.5 | -6.9 |
| 375 | -9.3 | 22.6 | 48.8 | 38.9 | 2.8 | -15.2 |
| Subtotal | 6.6 | 16.1 | 47.2 | 3.7 | 18.1 | 5.7 |
| Total | 3.1 | 13.9 | 33.0 | 6.8 | 13.2 | 6.6 |
| GROSS FIXED CAPITAL FORMATION | | | | | | |
| 310 | -30.7 | -10.9 | -100.3 | NA | 35.5 | -60.0 |
| 366 | 70.2 | 19.3 | 200.5 | -58.6 | 13.1 | 19.2 |
| 374 | 47.8 | 10.2 | -2.2 | 138.3 | 74.5 | -63.0 |
| 375 | 5.8 | 63.0 | -17.2 | 83.1 | 30.1 | -20.2 |
| Subtotal | 1.1 | 9.5 | -7.8 | 99.2 | 51.4 | -51.3 |
| Total | 6.4 | 24.4 | 0.2 | 73.0 | 33.2 | -44.5 |

NOTE: NIC 310 is tyre and tube industry. NIC 366 is electrical appliance industry. NIC 374 is automobile industry. NIC 375 is motorcycle industry.
SOURCE: Govt. of India, *Annual Survey of Industries* (various issues).

Consumer durable goods industries have developed constantly after economic reforms. Total GVA of consumer durable goods industry group at 1981-2 prices rose by 12.3 per cent per annum between 1991-2 and 1997-8. GVA of tyre and tube (NIC 310), electrical appliance (NIC 366), automobile (NIC 374) and motorcycles (NIC 375) industries increased rapidly (Table 1). Their relative contribution to growth of total GVA of the industry group between 1991-2 and 1997-8 were 13.7 per cent, 18.9 per cent, 30.7 and 11.0 per cent respectively. In other words, these four industries were leading consumer durable goods industries. As these industries cater mainly to the domestic market, exports are not important. GFCF rose to meet increasing domestic demand. Tyre and tube and automobile industries were the main contributors of GFCF between 1991-2 and 1997-8. These two industries together accounted for from 42.9 per cent to 62.9 per cent of the total GFCF of the industrial group during the same period. GFCF of automobile industry jumped up in 1995-6 and 1996-7 and pushed up the total GFCF of the industrial group. Many foreign automobile companies started production since 1994. Daewoo came in 1994, Honda in 1995, Hyundai in 1996, Toyota in

TABLE 2. GROWTH RATES OF GVA AND GFCF IN CONSUMER NON-DURABLE GOODS INDUSTRIES

(%)

| *Code* | *1992-3* | *1993-4* | *1994-5* | *1995-6* | *1996-7* | *1997-8* |
|---|---|---|---|---|---|---|
| GROSS VALUE ADDED | | | | | | |
| 20-1 | 1.2 | 23.2 | 19.2 | 2.2 | 15.1 | -5.5 |
| 247 | 42.8 | 14.0 | 3.2 | 15.9 | 65.2 | -43.5 |
| 304 | 15.3 | 23.9 | -4.3 | -1.5 | 42.7 | 8.2 |
| Subtotal | 16.5 | 20.1 | 8.2 | 5.8 | 38.1 | -18.6 |
| Total | 5.7 | 28.7 | 9.5 | -6.0 | 32.6 | -6.9 |
| GROSS FIXED CAPITAL FORMATION | | | | | | |
| 20-1 | 18.6 | -15.4 | 74.2 | 12.1 | -24.7 | -11.5 |
| 247 | -14.0 | 64.1 | 48.9 | 13.3 | -34.2 | 16.7 |
| 304 | 31.8 | 14.7 | 19.2 | 27.1 | 9.1 | -19.0 |
| Subtotal | 12.5 | 6.7 | 54.1 | 15.1 | -20.9 | -6.7 |
| Total | 19.9 | 16.6 | 40.1 | 17.9 | -9.9 | -8.1 |

NOTE: NIC 20-1 is food product industry. NIC 247 is man-made textile industry. NIC 304 is drug industry.

SOURCE: As in Table 1.

1997 and Ford in 1999. The development of automobile industry led the consumer durable goods industries in the 1990s. In electrical appliance industry, lumpy investment was implemented in 1994-5.

Consumer non-durable goods industries have also shown growth after the economic reforms. Total GVA of consumer non-durable goods industry group at 1981-2 prices rose by 13.2 per cent per annum between 1991-2 and 1996-7. GVA of food products (NIC 20-1), man-made textile (NIC 247), and drug (NIC 304) rose fast (Table 2). Their relative contribution to growth of total GVA of the industrial group between 1991-2 and 1996-7 were 22.2 per cent, 39.8 and 13.5 per cent respectively. Their investment also show a continued increase up to 1995-6. These three industries together accounted for from 41.3 per cent to 51 per cent of the total GFCF of the industrial group between 1991-2 and 1997-8. Development of food products after economic reforms is also noteworthy. Electrical appliance, automobile, motorcycle and man-made textile industries cater mainly to the high-income groups. But food products are wage goods. Among food product industries, sugar (NIC 206) and vegetable oil (NIC 211) industries have grown steadily untill 1996-7. Their GVA at 1981-2 prices rose by 11.2 per cent per annum and 24.5 per cent respectively between 1991-2 and 1996-7. The growth of food

TABLE 3. GROWTH RATES OF GVA AND GFCF IN INTERMEDIATE GOODS INDUSTRIES

(%)

| *Code* | *1992-3* | *1993-4* | *1994-5* | *1995-6* | *1996-7* | *1997-8* |
|---|---|---|---|---|---|---|
| GROSS VALUE ADDED | | | | | | |
| 301 | 15.2 | -13.0 | 31.7 | 31.2 | -14.5 | -35.6 |
| 302 | 137.9 | 28.3 | 13.6 | 139.8 | -28.3 | 3.2 |
| 306 | NA | 12.7 | -5.6 | 49.6 | -19.0 | 1.9 |
| 314 | 58.4 | 6.3 | 15.4 | 28.6 | -0.3 | -68.6 |
| 330 | 42.8 | 14.0 | 3.2 | 15.9 | -4.0 | 67.6 |
| Subtotal | 61.5 | 5.3 | 12.4 | 37.8 | -11.7 | -6.2 |
| Total | 25.5 | 6.2 | 12.9 | 31.5 | -8.3 | 1.5 |
| GROSS FIXED CAPITAL FORMATION | | | | | | |
| 301 | -25.9 | 142.4 | 11.2 | -25.5 | 18.4 | 25.2 |
| 302 | 247.3 | 12.5 | -44.5 | 1487.3 | -62.5 | 109.8 |
| 306 | 69.8 | -47.3 | 104.5 | 105.7 | -66.7 | 98.5 |
| 314 | 12.1 | 13.5 | -19.7 | -4.4 | 103.4 | -40.5 |
| 330 | -22.0 | 31.5 | 64.8 | -36.7 | -35.9 | -1.8 |
| Subtotal | -10.1 | 34.4 | 38.0 | 10.5 | -33.0 | 21.7 |
| Total | 15.9 | 21.3 | 35.3 | 12.2 | -8.0 | -6.1 |

NOTE: NIC 301 is fertilizer industry.
NIC 302 is plastic material and synthetic rubber industry.
NIC 306 is man-made fibre industry.
NIC 314 is refined petroleum product industry.
NIC 330 is iron and steel industry.

SOURCE: As in Table 1.

product industry suggests economic reforms had led to a trickle down effect. Change of consumer preference is reflected in the growth of man-made fibre industry. Consumer preference shifted from cotton to man-made fibre after 1976.[4] Exports of drugs have also increased rapidly during 1990s. Although the share of exports in output increased from 29.0 per cent in 1991-2 to 33.9 per cent in 1997-8.[5] The share of the domestic market is still much higher.

Because of backward linkage, total GVA of intermediate industry group at 1981-2 prices rose by 18.6 per cent between 1991-2 and 1995-6. GVA of fertilizer (NIC 301), plastic material and synthetic rubber (NIC 302), man-made fibre (NIC 306), refined petroleum products (NIC 314), and iron and steel (NIC 330) industries increased rapidly (Table 3). Their relative contribution to growth of total GVA of the industry group between 1991-2 and 1995-6 were 12.5 per cent, 14.6 per cent, 14.2 per cent, 13.3 per cent and 13.2 per cent respectively. Plastic material and synthetic rubber, man-made fibre,

TABLE 4. GROWTH RATES OF GVA AND GFCF IN CAPITAL GOODS INDUSTRIES

(%)

| Code | 1992-3 | 1993-4 | 1994-5 | 1995-6 | 1996-7 | 1997-8 |
|---|---|---|---|---|---|---|
| GROSS VALUE ADDED | | | | | | |
| 353 | -2.2 | 4.8 | 16.6 | 58.7 | -30.7 | -23.5 |
| 356 | 4.7 | 9.6 | 6.4 | 44.5 | -9.1 | -0.9 |
| 360 | 4.6 | -10.6 | 25.1 | 7.7 | -5.8 | 20.8 |
| 373 | 11.1 | -14.5 | 47.1 | 94.3 | -18.8 | -16.5 |
| Subtotal | 5.4 | -6.3 | 25.8 | 46.9 | -15.1 | -2.5 |
| Total | 10.3 | -4.5 | 19.7 | 36.7 | -2.3 | -5.9 |
| GROSS FIXED CAPITAL FORMATION | | | | | | |
| 353 | 62.6 | -8.4 | -13.1 | 48.6 | -52.8 | 140.2 |
| 356 | 112.1 | -75.5 | -9.7 | 148.4 | -23.4 | -15.0 |
| 360 | 14.0 | 54.2 | 79.1 | -21.0 | -17.8 | -43.0 |
| 373 | 35.7 | 49.0 | -50.7 | 211.4 | -15.2 | -43.6 |
| Subtotal | 59.4 | -11.3 | -7.8 | 69.0 | -21.2 | -27.2 |
| Total | 48.7 | -4.3 | 17.8 | 42.0 | -4.9 | -33.2 |

NOTE: NIC 353 is industrial machinery for food and textile industry.
NIC 356 is non-electrical machinery industry.
NIC 360 is electrical industrial machinery industry.
NIC 373 is heavy motor vehicle industry.

SOURCE: As in Table 1.

refined petroleum product and iron and steel industries had lumpy investment in 1995-6, 1995-6, 1996-7 and 1994-5 respectively. Although GFCF of each industry has fluctuated, subtotal of five industries have grown constantly from 1991-2 to 1995-6. These five industries accounted for from 37.9 per cent to 60.3 per cent of total GFCF of the industry group between 1991-2 and 1997-8. The development of automobile industry expanded the demand for intermediate goods. Synthetic rubber industry supplies material to tyre and tube industries, which are classified under consumer durable goods industry. GVA of tyre and tube industries at 1981-2 prices rose by 11.2 per cent per annum between 1991-2 and 1997-8. Moreover, automobile industry also created the demand for iron and steel. In 1989-90, Leontief multiplier of motor vehicle on iron, steel and ferro alloys was 0.087461. It was the third highest among 115 commodities after motor vehicles and trade. Rapid growth of automobile industry has induced rise of iron and steel industry.

Investment boom had created demand for capital goods industry in the early 1990s. Total GVA of capital goods industry group at

1981-2 prices rose by 14.6 per cent between 1991-2 and 1995-6. GVA of industrial machinery for food and textile industry (NIC 353), non-electrical machinery (NIC 356), electrical industrial machinery (NIC 360), and heavy motor vehicle (NIC 373) industries came up fast (Table 4). Their relative contribution to growth of total GVA of the industry group between 1991-2 and 1995-6 were 8.1 per cent, 8.9 per cent, 7.2 per cent, and 30.6 per cent respectively. Non-electrical machinery, electrical industrial machinery, and heavy motor vehicle industries had lumpy investment in 1992-3, 1994-5 and 1995-6 respectively. As mentioned before, growth of food product and man-made textile industries induced investment in capital goods industries. As a result, GVA of industrial machinery for food and textile industry rose substantially in 1995-6. GVA of heavy motor vehicles also jumped up in 1995-6. It can be estimated from the phenomena that transport of volume increased in the early 1990s. Production of medium and heavy commercial vehicles pushed up GVA of capital goods industry.

The coming of age of the stock market also might have encouraged investment in the 1990s. Since 1991, reforms were started in the credit and capital markets. In 1992 the Securities and Exchange Board of India (SEBI) Act replaced the Capital Issues Act. Prior to 1992, the government very closely regulated the primary issues market in almost every aspect. Now companies were free to approach the capital market after they got a clearance from SEBI. Total amount of new capital issues by non-government public limited companies including preference shares and debentures at 1981-2 prices jumped up from Rs 2,927 crore in 1991-2 to Rs 8,525 crore in 1992-3.[7] It reached Rs 9,875 crore in 1994-5.[8] Number of issues as well as stock prices increased between 1991-2 and 1994-5. Annual average price earning ratio rose from 26.05 in 1991-2 to 41.24 in 1994-5.[9] The stock market boom was promoted by not only relaxation of the regulation but also with the inflow of foreign portfolio investment (FPI). Quarterly net inflow of FPI had been more than US $900 million between the third quarters of 1993-4 and 1994-5.[10] During the period, monthly average of price earning ratio had been above 32. However, the boom was over in 1994-5.

The stock boom was created by a small number of companies with strong performances. GFCF also has the same character. If a big company has lumpy investment, it might push up GFCF of the

TABLE 5. AVERAGE SHARE OF PROFIT AND DEPRECIATION IN GROSS INCOME

(%)

| | *1981-2 to 1990-1* | *1991-2 to 1997-8* |
|---|---|---|
| Capital goods | 37.3 | 40.3 |
| Intermediate goods | 52.9 | 60.5 |
| Consumer durable goods | 40.9 | 52.5 |
| Consumer non-durable goods | 38.3 | 51.5 |

SOURCE: As in Table 1.

industry. The private sector got alternative source of funds in addition to loan from financial institutions. In fact, the survey on financial performance of non-government non-financial public limited companies by RBI shows that the share of paid-up capital in source of funds rose suddenly from 7.1 per cent in 1992-3 to 29.6 per cent in 1993-4 and 26.8 per cent in 1994-5.[11] The private corporate sector adjusted itself to the rise of interest rate after financial liberalization in 1991.[12]

In order to analyse internal financing, the share of profits in gross income is to be calculated. As the net income in ASI is obtained by deducting the rent paid and interest paid from the net value added, net income represents the factor share of employees and the entrepreneur in the net value added. Profits are excess of net income over the cost of employees' compensation, i.e. total emoluments and supplements to emoluments, i.e. (1) contribution to provident and other funds, and (2) workmen and staff welfare expenses. Depreciation is a source of investment. If the share of profit and depreciation in gross income, i.e. net income plus depreciation, increases, firms can allot higher percentage of income for investment. Table 5 shows average share in the 1980s and 1990s. In the four industry groups, the share rose in the 1990s. One can conclude that the share of employees' compensation cost declined in the four industry groups during the 1990s. Companies could now use internal financing to diversify their production without expanding their capacity. But an increase in internal financing is also advantageous to investment.

In investigating the relative price of capital goods vis-à-vis products, capital goods index is divided by chain index of industry group (1981-2 = 100). When the ratio decreases, the relative prices of capital goods becomes cheaper. Investment is encouraged by the cheaper prices of capital goods. The ratio fell from 1.000 and 1.097 in 1991-2 to 0.932 and 0.989 in 1995-6 in intermediate and consumer non-durable goods industries. Decrease of the relative price of capital goods is advantageous to investment in intermediate and consumer non-durable goods industries. Although the relative prices in consumer durable goods industries rose, investment increased in the first of 1990s.[13] It seems that the relative prices of capital goods do not affect investment.

The mid-1990s witnessed lumpy investment in the industries where GVA had grown. The aggregated picture of investment in each industry group showed rapid growth of GFCF in the first half of 1990s, which created demand for capital goods industry.

## 2. THE END OF INVESTMENT BOOM

The investment boom in India started in the first half of 1990 and was over by mid-1990s. GFCF declined in capital goods, intermediate goods, consumer non-durable goods and total industries after 1995-6 and in consumer durable goods industry after 1996-7. Decrease of GFCF was reflected in the decline of GVA in capital goods industry. As some industries had lumpy investment even after 1995-6, GVA of capital goods industry in 1996-7 and 1997-8 was still much higher than GVA in early 1990s. But one cannot expect rapid growth of GVA of capital goods industries. Five reasons could be attributed to the end of the investment boom. First, demand declined or stagnated. Second, as new capital issues by private corporate sector dropped, firms could not procure funds. Third, tightened monetary policy in 1995-6 refrained investment. Fourth, payment of high interest became burden of manufacturers. Fifth, imports substituted domestic production. Further, an increase in imports too, may have affected domestic production. These five reasons will now be examined.

Although high growth rates of demand encouraged investment in the first half of 1990s, GVA started declining in the mid-1990s. If there was a decrease or stagnation in demand and fixed capital stock increased continuously, then underutilization of capacity would have occurred. This study estimates utilization ratio on the basis of minimum capital output ratio.[14] Fixed capital output ratios are calculated on the 1981-2 prices. A benchmark year is then selected on the basis of the observed lowest capital output rate in the 1990s. Since market conditions have changed after economic reforms, benchmark years are selected from the 1990s. The lowest observed capacity output ratio is considered as capacity output. The estimate of capacity is obtained by dividing the real fixed capital stock by the minimum capital output ratio. The utilization rate is given by actual output as a proportion of the estimated capacity.

Thus,

$$U = \frac{O}{\underline{C}} \times 100$$

$$\underline{C} = \frac{C}{(C/O)min}$$

where U is capacity utilization, O is GVA, $\underline{C}$ is estimate of capacity, C is gross fixed capital stock (GFCS).

GFCS was estimated by the perpetual inventory method. A GFCS in year T, denoted by $K_T$, is based on the following formula:

$$K_T = K_o + \sum_{t=1}^{T} (I_t - DS_t)$$

where $K_o$ is fixed capital stock at constant price in the benchmark year, $I_t$ is the gross investment at constant price in fixed assets in year t, and $DS_t$ is the amount of fixed assets at constant price discarded during year t. $I_t$ is defined as

$$I_t = F_t / P_t$$

where $F_t$ is GFCF and $P_t$ is the capital goods price. It is difficult, however, to obtain data of $DS_t$. In this study it is assumed that $DS_t$ is zero. According to A. Banerji doubling of the book value of the benchmark year serves as a measure of replacement value of fixed assets for the year.[15] But Goldar maintains that the choice of benchmark year is arbitrary.[16] Nevertheless, the present study uses the Banerji model. Goldar, Hashim and Dadi estimated the replacement value of fixed assets in benchmark years after analysing the balance sheets of firms.[17] The benchmark years 1951 and 1960 respectively selected by them were outdated because the industrial structure has changed in the last four decades. The purpose of this study is to estimate the underutilization of capacity. The year 1978-9 is selected as the benchmark year because GFCF data is available only after 1979-80. The trends of GFCS are more important than its absolute level for this purpose.[18] If the perpetual inventory method takes longer period, the weight of margins of errors in benchmark year becomes less.

TABLE 6. GROWTH RATES OF CAPITAL AND LABOUR PRODUCTIVTY DURING THE 1990s

(%)

| | *Pre-peak year* | | *Peak year* | *Post-peak year* | |
|---|---|---|---|---|---|
| | *Capital* | *Labour* | | *Capital* | *Labour* |
| Capital goods | 5.3 | 9.1 | 1995-6 | -9.7 | -2.3 |
| Intermediate goods | 7.0 | 11.5 | 1995-6 | -9.6 | -0.1 |
| Consumer non-durable goods | 5.7 | 11.7 | 1993-4 | -4.5 | 2.9 |
| Consumer durable goods | 4.8 | 11.8 | 1994-5 | -3.6 | 5.3 |
| Total | 3.7 | 9.0 | 1995-6 | -6.4 | 2.1 |

SOURCE: As in Table 2.

Although utilization rates of the four industry groups peaked in the mid-1990s, they witnessed a clear drop after the mid-1990s. Therefore, it can be concluded that underutilization of capacity made its appearance across the four industrial groups. To confirm this conclusion, capital productivity (GVA per GFCS) and labour productivity (GVA per worker) during pre-peak year and post-peak year are calculated. Compound growth rates of capital and labour productivity dropped during the period of the post-peak year (Table 6). Particularly in the capital and intermediate goods industries, labour productivity became negative during the post-peak period. The trends of labour productivity can be explained by the utilization rate of capacity. In other words, capital and labour productivity went down due to underutilization of capacity. Although GVA started to decline after 1996-7 in consumer non-durable goods industry group and GVA had risen constantly in consumer durable goods industry group, utilization rates dropped after 1993-4 and 1994-5 respectively. GFCF of consumer non-durable and durable goods industries declined with two years lag after utilization rates started to decrease. Decline of utilization reduced their GFCF at 1981-2 prices with two years lag. It is noteworthy that underutilization made its appearance in consumer durable and non-durable goods industries earlier than in capital and intermediate goods industries. When producers of consumer non-durable and durable goods show decreased rate of growth of production, it naturally affects capital and intermediate goods industries. Annual growth rates of material consumed in consumer non-durable and durable goods industries dropped from 9.6 and 20.4 per cent between 1991-2 and 1995-6 to 1.3 and 2.5 per cent between

1995-6 and 1997-8 respectively. This restricted the demand of the intermediate goods industry. The utilization rates of the intermediate goods industry group dropped after 1995-6 and its GFCF also declined after 1995-6 without any time lag; the reason being lumpy investment by efficient and big companies. Expansion of capacity by big companies increased output sharply. However, if demand grew by lower rate than output, over-capacity might have occurred immediately.

Sudden drop of new capital issues might have made it difficult to procure funds from stock market. The total amount of new capital issues by non-government public limited companies including preference shares and debentures at 1981-2 prices went down sharply from Rs 9,875 crore in 1994-5 to Rs 1,028 crore in 1997-8. On the other hand, amounts disbursed by all financial institutions at 1981-2 prices jumped from Rs 12,549 crore in 1994-5 to Rs 17,578 crore in 1997-8.[19] In other words, industries could procure funds from financial institution. The reasons behind the decrease of new capital issue could be ascribed to two reasons. First, the lending rates of financial institutions fell. In India, financial institutions supply long-term funds for investments. The prime lending rates of IDBI (Industrial Development Bank of India) went down from 16-19 per cent in 1995-6 to 13.3 per cent in 1997-8. Second, there was a decline in stock prices. The annual average of price earning ratio dropped from 41.24 in 1994-5 to 19.92 in 1995-6. Therefore, attractiveness to stock issue was adversely affected by this decline of the price earning ratio. A survey of the non-governmental, non-financial public limited companies shows a decrease in the share of paid-up capital as sources of funds, from 26.8 per cent in 1994-5 to 8.3 per cent in 1997-8. On the other hand, the share of loans from financial institutions rose from 3.9 per cent to 9.9 per cent during the same period. There is a substitutive function between the stock market and financial intermediaries in terms of financing private investment.[20] Although fluctuation of lending rates causes shift of financial source, this is not reflected in trends of GFCF clearly.

When the RBI tightened its monetary policy in 1995-6, interest rates of financial institutions rose. Interest rates are correlated with the disbursements of financial institutions. Disbursement by all the financial institutions which was deflated at 1981-2 prices came up by 5.4 per cent in 1995-6 when the prime lending rate of the IDBI was raised from 15 per cent to 16-19 per cent. On the other hand,

TABLE 7. GROWTH RATES OF WORKING CAPITAL AND INTEREST PAID AT 1981-2 PRICES BETWEEN 1995-6 AND 1997-8

(%)

| | *GVA* | *Working capital* | *Interest paid* | *Profits* |
|---|---|---|---|---|
| Capital goods | -3.3 | 16.5 | 3.8 | -18.0 |
| Intermediate goods | -1.8 | 14.6 | 13.5 | -20.9 |
| Consumer non-durability | 7.3 | 9.5 | 9.0 | 7.2 |
| Consumer durability | 9.1 | 55.4 | 2.1 | 12.0 |

SOURCE: As in Table 1.

disbursements increased by 22.8 per cent in 1997-8 when lending rates were reduced from 16.2 per cent to 13.3 per cent. A sample survey by the IDBI shows that the total long-term borrowing of 520 companies at 1981-2 prices went down by 14.1 per cent but total GFCF at 1981-2 prices came up by 14.5 per cent in 1995-6.[21] Another sample survey by IDBI shows that the total long-term borrowing of 550 companies rose by 13.6 per cent but the total GFCF decreased by 3.1 per cent due to inventory increase by 28.6 per cent in 1997-8.[22] As investment strategy is a long-term derivative it is not affected immediately by change of interest rates.[23]

Under the circumstances that GVA was stagnating or declining after 1995-96, payment of interest became burden for manufacturers. Although they had borrowed long term loan for investment, their investment created overcapacity. While they could not recover investment, they had to pay interest. Moreover, decline of GVA made period of capital turnover longer. They might need more working capital to produce the same output if input output ratio is same. In spite of decline of GVA, therefore, working capital of capital goods and intermediate goods industries at 1981-2 prices grew by 16.5 per cent and 14.6 per cent per annum respectively (Table 7). Consequently, interest for short and long term loan paid by capital goods and intermediate goods industries at 1981-2 prices came up by 3.8 and 13.5 per cent per annum respectively. Decrease of profits restricted investment. High interest above ten percent was not a direct cause of decline of GFCF. But high interest deteriorated the situation in these two industry groups.

Total imports in terms of dollars increased by 13.5 per cent per annum between 1991-2 and 1997-8. This can be ascribed to three reasons. First, import licensing was relaxed after economic reforms. Second, import duty was reduced after economic reforms. Third, domestic demand expanded in the first half of 1990s. On the other hand, rupee rates vis-à-vis one US dollar were depreciated from 31.5 in August 1995 to 39.5 in March 1998 which might have restricted imports. Reduction of import duty is offset by depreciation of rupee.

Trends of imports depend on domestic demand. Imports of capital goods deflated by 1981-2 prices rose till the mid-1990s but stagnated since 1995-6 (Table 8). Although imports of machinery (barring electric and electronic) rose rapidly till 1995-6, they went down after 1995-6. Imports of machinery accounted for from 32.8 per cent to 43.6 per cent of the total capital goods imports in the 1990s. On the other hand, imports of electronic goods rose rapidly. Investment boom induced the import of machinery till 1995-6. Sudden depreciation of the rupee after September 1995 was advantageous to domestic manufacturers. But the production of capital goods grew till 1995-6 and stagnated after 1995-6. Capital goods industries consists of 13 industries. In all these industries except NIC 378 (bullock carts and pushcarts), production rose up to mid-1990s. The degree of dependence on imports is different amongst different industries. We can calculate the share of import in domestic consumption, i.e. total output plus import from the *Input-Output Transactions Table*. The

TABLE 8. GROWTH RATES OF CAPITAL GOODS IMPORTS

(%)

| | *1992-3* | *1993-4* | *1994-5* | *1995-6* | *1996-7* | *1997-8* | *1998-9* |
|---|---|---|---|---|---|---|---|
| Manufacture of metals | 18.7 | 28.4 | 4.8 | 33.5 | 15.7 | 6.1 | 30.0 |
| Machine tools | 2.0 | -2.3 | -30.8 | 211.9 | 43.6 | -17.1 | -8.6 |
| Machinery except electrical and electronic | 20.3 | 19.5 | 31.4 | 42.4 | -5.5 | 2.5 | -6.5 |
| Electrical machinery except electronic | 39.3* | -74.1* | 11.6 | 52.1 | -14.3 | 19.9 | 23.8 |
| Electronic goods | N.A. | N.A. | 22.0 | 41.3 | -17.3 | 51.3 | 18.4 |
| Computer goods | N.A. | N.A. | 119.5 | 174.2 | -28.9 | 107.7 | 6.3 |
| Transport equipment | 32.0 | 188.8 | -20.6 | -1.8 | 36.6 | -26.9 | -15.5 |
| Project goods | -7.8 | 33.3 | 3.5 | 27.7 | -9.9 | -15.3 | 71.9 |
| Total | 25.8 | 49.2 | 20.9 | 45.9 | 1.9 | 3.4 | 16.3 |
| | (6300.9) | (9400.4) | (11369.5) | (16588.7) | (16909.7) | (17478.2) | (20327.1) |

NOTE: Figures in parenthesis indicate total import in 1981-2 prices, which are Rs in crore. It is deflated by wholesale price index of machinery and machine tools.
* The figure include electronic goods and computer goods.

SOURCE: As in Table 1.

share in industrial machinery for food and textiles, industrial machinery for others, machine tools, electric industrial machinery, and motor vehicles was 18.2 per cent, 7.7 per cent, 22.3 per cent, 6.6 per cent and 3.6 per cent respectively in 1989-90.[24] We have more detailed examination in the next paragraph.

## 3. CAPITAL GOODS INDUSTRIES IN THE 1990s

The effect of investment boom is reflected in the production and investment of capital goods industries. As mentioned before, NIC 353, 356, 360 and 373 are playing important roles in capital goods industry. We will examine them turn by turn. The machine tool industry is investigated in detail to analyse the impact of imports.

### *(1) Industrial Machinery for Food and Textile Industries (NIC 353)*

Growth of investment in food and man-made fibre industries created demand for this industry. GVA at 1981-2 prices rose up to 1995-6 and declined after 1995-6. GFCF at 1981-2 prices came up in the 1990s to meet rising demand. Over-capacity occurred in 1995-6. In spite of that, there was lumpy investment in 1997-8. Growth rates of capital and labour productivity declined by 11.2 per cent per annum and 8.6 per cent from 1991-2 to 1993-4, rose by 25.6 per cent and 33.8 per cent from 1993-4 to 1995-6 and dropped by 31.2 per cent and 19.3 per cent from 1995-6 to 1997-8 respectively. Material consumed at 1981-2 prices followed the same trends. It went down by 8.5 per cent, increased by 30.4 per cent and fell by 9.9 per cent during the same period. The figures for the number of workers also showed the same trends. It increased from 1991-2 to 1995-6 and fell after 1995-6 onwards.

Imports of industrial machinery for food and textiles followes the same trends along with the domestic output. Imports of industrial machinery for food (Indian Trade Classification 8433, 8434, 8435, 8437, and 8438) and industrial machinery for textiles (ITC 8444, 8445, 8446, 8447, and 8448) increased by 39.8 per cent per annum and 58.5 per cent respectively, between 1991-2 and 1995-6 but decreased by 13.1 per cent and 16.5 per cent respectively, between

1995-6 and 1998-9.[25] The share of imports in domestic consumption, i.e. output plus imports minus exports in industrial machinery for food and textiles industries rose from 23.9 per cent in 1992-3 to 36.8 per cent in 1994-5 but fell to 27.2 per cent in 1997-8.[26] Therefore, it seems that domestic production was not affected by imports.

*(2) General Purpose Non-electrical Machinery (NIC 356)*

This category includes lifting and handling equipment, pumps compressors, taps and valves and bearing, gears, etc. Although GVA at 1981-2 prices came up constantly by 15.2 per cent per annum between 1991-2 and 1995-6, it declined after 1995-6 onwards. On the other hand, there was lumpy investment in 1992-3 and 1995-6. Consequently, over-capacity occurred and utilization rates went down after 1995-6. Growth rates of capital and labour productivity increased by 4.2 per cent per annum and 9.9 per cent between 1991-2 and 1995-6 and decreased by 10.6 per cent and 6.0 per cent between 1995-6 and 1997-8 respectively. Material consumed at 1981-2 prices rose by 9.7 per cent and fell by 6.9 per cent during the same period. Number of workers increased by 4.9 per cent from 1991-2 to 1995-6 and by 1.0 per cent from 1995-6 to 1997-8.

*(3) Electrical Industrial Machinery (NIC 360)*

This industry shows different picture. GVA at 1981-2 prices rose by 10.3 per annum between 1981-2 and 1990-91. As demand increased, investment came up to meet it. GFCS at 1981-2 prices grew by 6.5 per cent during the same period. As GFCF accelerated after 1992-3, GFCS at 1981-2 prices increased by 7.0 per cent from 1990-1 to 1997-8. But annual growth rates of GVA at 1981-2 prices fell to 3.9 per cent during the same period. As a result, utilization of capacity went down from 1990-1. Growth rates of capital and labour productivity decreased from 3.6 per cent and 8.2 per cent between 1981-2 and 1990-1 to -2.9 per cent and 1.9 per cent between 1990-1 and 1997-8 respectively. Growth rates of material consumed in 1981-2 prices went down from 9.2 per cent between 1981-2 and 1992-3 to 2.8 per cent between 1990-1 and 1997-8. Number of workers increased constantly during the 1980s and 1990s.

*(4) Automobile Industry (NIC 373 and 374)*

ASI changed classification in 1989-90. NIC 374 in 1970 code was divided into NIC 373 (heavy motor vehicle) and NIC 374 (motor cars). Data on NIC 373 can be obtained after 1989-90 onwards. GVA at 1981-2 prices was stagnant between 1989-90 and 1993-4, grew rapidly between 1993-4 and 1995-6 and fell after 1995-6 onwards. Material consumed at 1981-2 prices also followed the same trends. The industry witnessed lumpy investment in 1995-6 and 1996-7. Company-wise data on production of buses and trucks are available. TELCO, Ashok Layland, and Hindustan Motors produce trucks. Production of TELCO rose by 39.0 per cent and 21.8 per cent in 1995-6 and 1996-7 respectively.[27] In other words, large-scale investment by TELCO expanded capacity and raised output. But production of trucks suddenly dropped in 1997-8.

To calculate GFCS, we need data for the 1980s. We analyse combined data of NIC 373 and 374, i.e. NIC 374 in 1970 code. The share of NIC 373 and 374 in net value added was 55.5 per cent and 45.5 per cent in 1989-90. Investment in automobile industry accelerated in the 1990s. GFCS jumped up by 20.6 per cent in 1995-6 and by 22.7 per cent in 1996-7. Although GVA at 1981-2 prices rose up to 1995-6, it has stagnated since 1995-6 onwards. As a result, utilization rate went down after 1995-6. Capital and labour productivity increased by 17.1 per cent and 18.2 per cent between 1991-2 and 1995-6 and decreased by 13.7 per cent and 6.5 per cent between 1995-6 and 1997-8 respectively.

*(5) Machine Tools (NIC 357)*

Machine tools are power driven machines used for cutting, forming, shaping or processing metals or other materials into desired forms. Since these materials are used for the manufacture of machine tools as well as other machines, the development of the machine tools industry indicates the technological level of the country and the condition of the machine industry as a whole. Although the machine tools industry accounted for only 2.3 per cent of total GVA of capital goods industry group in 1997-8, this sector is crucial for technological development. If many units had introduced new technology, the demand for machine tools might have shifted from ordinary machine

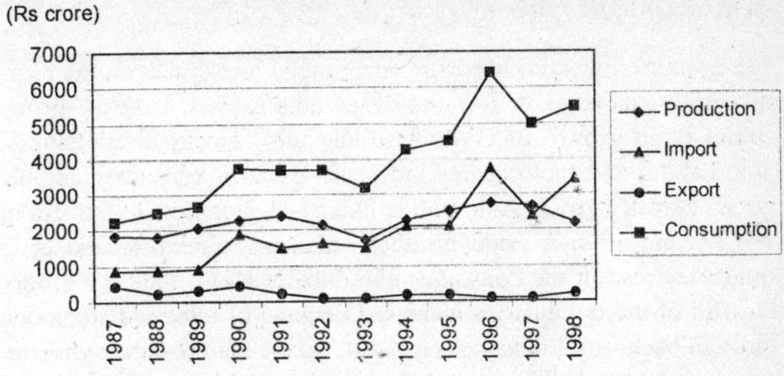


FIGURE 3. PRODUCTION OF MACHINE TOOLS AT 1981-2 PRICES

tools to sophisticated ones. Therefore, impact of economic reforms is reflected in production and import composition.

NC (numerical control) machine tools diffused in the second half of 1980s. The share of NC machine in total production of machine tools rose from 8.5 per cent in 1986 to 31.1 per cent in 1989. The share came up to 37.3 in 1993. As non-NC machine was substituted by NC machine, production of automatics and non-NC lathe machine declined.[28] However, the share has not changed after 1994. Although the production of automatics declined, that of non-NC lathe machine remained at the same level.

Figure 3 shows that the trends of consumption of machine tools depends on imports. Domestic production kept up at the same level as the pre-reform period. Imports of machine tools rose after the economic reforms. Consequently, the share of domestic production in domestic consumption fell from 65.3 per cent in 1991 to 40.4 per cent in 1998. Domestic demand expanded because automobile and automobile component industries developed. Construction of new automobile plants increased imports of machine tools. Hyundai and TELCO plants were set up in 1996 and 1998, respectively. When new automobile plants were set up, imports exceeded production. In 1998-9 imports by TELCO accounted for 19.7 per cent of total imports of machine tools.[29] Domestic manufacturers do not have capacity to cater for demand from lumpy investment. Although the share of domestic production in consumption went down, it was not substituted by imports on a large-scale.

## CONCLUSION

Abolition of industrial licensing encouraged investment on the basis of entrepreneurship. A few industries among each industry group shows rapid growth of GVA. Tyre and tube, electrical appliances, automobile, and motorcycles industries grew in consumer durable goods industry group. Automobile industry had lumpy investment in 1995-6 and 1996-7. Food products, man-made textiles, and drug industries rose in the consumer non-durable goods industry group. Growth of these industries increased demand of intermediate goods through backward linkage. As a result, plastic material and synthetic rubber, man-made fibre, refined petroleum products, and iron and steel led to the intermediate goods industries. To meet the expansion in demand, lumpy investment was implemented in the four industries between 1994-5 and 1996-7. Large-scale investment created demand for capital goods. GVA of capital goods industry group accelerated after 1991-2.

Production capacity grew faster than demand. Consequently, over-capacity made appearance and utilization of capacity declined. GFCF decreased with time lag. Although GVA of capital goods industry group in 1996-7 and 1997-8 was still much higher than GVA in early 1990s, we cannot expect its rapid growth. Stock boom and increase of internal financing might have generated advantageous circumstances for investment. But these two reasons cannot explain why GFCF declined after 1995-6. Tightening policy of RBI in 1995-6 might have had an impact on the macro-economy. As investment is a long-term derivative, change of interest rate does not affect GFCF directly. Only decline of utilization can explain reduction of GFCF after 1995-6.

Imports of capital goods depend on domestic demand. Their imports increased between 1991-2 and 1995-6. During this period, GVA of all capital goods industries except bullock and pushcarts rose. After 1995-6 both imports and domestic production stagnated. In machine tool industry the share of production in domestic consumption fell to 65.3 per cent in 1991 and 40.4 per cent in 1998. When new automobile plants were set up, imports exceeded production. Domestic production was maintained at the same level in the 1990s and it has not been substituted on a large-scale.

## NOTES

1. C.P. Chandrasekhar, 'Explaining Post-Reform Industrial Growth', *Economic and Political Weekly*, Special Number, 1996, p. 2543.
2. Govt. of India, *National Accounts Statistics: Factor Incomes 1980-1–1989-90*, 1994.
3. In investigating the issue of whether there was acceleration in the growth rate of GFCF after economic reforms, dummy variables are used to allow both the intercept and the slope to be different after 1991 onwards. Thus we estimate a typical regression equation:

$$\ln Y = \alpha + \alpha'D + \beta t + \beta'Dt$$

where $\alpha$ is a dummy variable with values equal to zero for the years up to 1990-1 and one thereafter. The coefficient of the multiplicative dummy term, $\beta'$ if it is positive and significant, shows that there was acceleration in the gross rate in the post-economic reform period.

4. S. Uchikawa, *Indian Textile Industry: State Policy, Liberalization and Growth*, New Delhi: Manohar, 1998, p. 77.
5. Homepage of Ministry of Chemical and Fertilizers, http://www.nic.in/cpc/pharma.htm.
6. Govt. of India, *Input-output Transactions Table 1989-90*, 1997, p. 254.
7. The amount is deflated by chain index of capital goods.
8. RBI, *Handbook of Statistics on Indian Economy*, 2000, p. 95.
9. Ibid., p. 425.
10. Altough net inflow of FPI continued up to 1998, it could not sustain stock market. Ibid., pp. 302-3.
11. RBI, *RBI Bulletin* (various issues).
12. K. Sen and R.R. Vaidya, *The Process of Financial Liberalization in India*, New Delhi: Oxford University Press, 1997, p. 130.
13. Ramaswamy points out real prices of consumer durable goods, relative to all manufacturing goods, have declined in the post-reform period under competitive conditions in Chapter five.
14. D.U. Sastry, *The Cotton Mill Industry in India*, New Delhi: Oxford University Press, p. 28.
15. A. Banerji, *Capital Intensity and Productivity in Indian Industry*, New Delhi: Macmillan, 1975, p. 21.
16. B.N. Goldar, *Productivity Growth in Indian Industry*, New Delhi: Allied, 1985, p. 58.
17. S.R. Hashim and M.M. Dadi, *Capital-Output Relations in Indian Manufacturing: 1946-1964*, Baroda: MS University of Baroda, 1973, p. 13.
18. Goldar pointed out the difference in TFP estimates between studies may be attributed largely to the differences in capital estimates. B.N. Goldar, 1985, p. 50.
19. RBI, 2000, p. 87.
20. M. Nagaishi, 'Stock Market Development and Economic Growth: Dubious Relationship', *Economic and Political Weekly*, 17 July 1999, p. 2011.

21. IDBI, *Financial Performance of IDBI Assisted Companies in the Private Corporate Sector*, 1995-6, pp. 22-3.
22. IDBI, 1998-9, pp. 22-3.
23. We cannot find out relation between long-term borrowing and GFCF with one year time lag.
24. The share of export of industrial machinery for food and textile, industrial machinery for others, machine tools, and electric industrial machinery in total output was 13.4 per cent, 12.1 per cent, 5.2 per cent, and 1.6 per cent respectively in 1989-90. As weight of exports of capital goods in total output has been low, this chapter does not examine exports (Govt. of India, *Input-Output Transactions Table 1989-90*, 1997, p. 226).
25. Govt. of India, *Monthly Statistics of Foreign Trade of India* (various issues).
26. As output do not include unorganized sector, this share is overestimated.
27. Automotive Component Manufacturers Association of India, *Facts and Figure*, 1999-2000, p. 19.
28. S. Uchikawa, 'Economic Reforms and Foreign Trade Policies: Case Study of Apparel and Machine Tools Industry', *Economic and Political Weekly*, 27 November 1999, p. M147.
29. Indian Machine Tool Manufacturers' Association, Data on Import of Machine Tools, 1998-9.

# Intermediate Goods Industries: A Case Study of the Cement and Steel Industries

RAJENDRA R. VAIDYA

## I. INTRODUCTION

The reform process begun in 1990-1 has had a profound impact on the working of the industrial sector. As is well known the reform touched industry, trade and financial sectors. The impact of all these reforms is currently being debated in the country. In what follows we concentrate on the intermediate goods with a special emphasis on cement and steel.

These two industries have been chosen, first because both have been completely decontrolled while other important intermediate goods like fertilizers and petroleum are yet to be deregulated. Second, broadly speaking both can be called homogenous goods, in the sense that consumers would not be primarily driven in their purchase by the identity of the producer. Though some cement and steel firms have made attempts to brand their products we assume that the overall impact of these efforts to decrease substitutability between output of rival firms is insignificant. Given the nature of both the products this does not seem to be an unreasonable assumption. This is important because if there is reason to believe that substantial product differentiation has been achieved by firms, then inter-firm differences in all financial performance and market structure measures need to be interpreted with caution. In the case of reasonably homogenous products inter-firm differences have a rather straightforward interpretation.

Though quiet similar in the above two ways, both these industries differ substantially in other important aspects. Cement is a very thinly traded commodity in the world and its imports have always been negligible and its export is possible only to neighbouring countries because of the high transportation costs involved. Steel, on the other hand is a very important commodity in world trade and India has always participated in this trade both as an importer and an exporter. Scale economies are much more pronounced (in the sense that a plant of minimum efficient scale would have a capacity that would be a large fraction of the total demand for the product) in the case of steel than cement. Though for cement too, scale economies are important.

In what follows we attempt to ascertain the impact that reforms have had on the steel and cement industry at both the level of the industry as a whole and at the firm specific level. At the level of the industry (or commodity) some of the important aspects of the possible effects of reforms are as follows:

(a) The behaviour of the rate of growth of production.
(b) Fall in the real price of the product as a consequence of greater domestic production and/or greater availability of imports. The difference on the rates of growth of price level of the commodity and the rate of growth of the manufactured goods as a whole would reflect this.
(c) The rise in productivity that would be experienced as a consequence of a possible rise in investment. This would be reflected by variables like labour productivity.
(d) From the perspective of workers in the sector the rise in productivity would increase wages. This would be reflected in the behaviour of wages per worker.

Our main interest in looking at the impact of reforms on these two industries at the firm level is as follows. One would expect market shares of leading firms in particular and concentration ratios in general to fall in the post-reform period, with obvious consequences on the scope of firms to abuse monopoly positions. The entire reform package of various sectors is expected to have a generally favourable effect in terms of investment and the advantages that come with greater freedom that firms would have in their operations. This freedom is likely to lead to possible diversification of activity by firms and attempts to penetrate international markets. These advantages would have to be weighed against the costs that such a reform package would impose. When a tightly structured control regime (both in terms of domestic controls on productive activity and international trade of the commodity in question) is dismantled this would automatically lead to some financial distress among firms. Furthermore if financial sector reforms lead to high interest rates this too would put pressure on firm performance both in terms of current profits and investments. Financial sector reforms have provided firms with enormous flexibility in their financing patterns. All the major sources of external funds, (consisting of the capital market, financial institutions and banks) have been allowed much greater freedom of operations than in the pre-reform era (for details regarding financial liberalization see Sen and Vaidya (1997). If reforms on the whole lead to a large number of financially weak firms within particular industries, then this would be a matter of great concern, especially if the financial institutions have a large exposure in that sector.

In the current context, with regard to the Industrial Development Bank of India (IDBI), the total outstanding debt as of March 1999 was about 13 per cent to the steel industry and about 5 per cent was to the cement industry. By any standards this is a substantially large exposure. These two industries also accounted for about 16 per cent of total non-performing assets in March 1999. In the case of Industrial Credit and Investment Corporation of India (ICICI) both these numbers are slightly lower. The important implication here is that if a large number of firms who have borrowed funds from these institutions experience losses then this would have a negative impact on these financial institutions given their large exposure to these industries. This would force the government to rethink its strategy of liberalization on both the real and financial fronts, if it is perceived that the costs imposed by liberalization in the short run far exceed a critical level which the government finds acceptable. What complicates the issue further is that there is no way in which the negative effects and costs of liberalization can be foreseen and planned for. Thus the only options open for the government is to liberalize and wait and watch its effects. If the results are extremely unfavourable to a particular interest group (consisting broadly of consumers, producers and financial institutions) then in a democratic set-up a rethinking of liberalization strategies (and in an extreme case a complete rollback of reforms) is inevitable.

In the case of consumers if reforms lead to a lowering of prices this is undoubtedly a good thing for them. If on the other hand prices rise after liberalization, consumers being a powerful interest group could force the government to retrace its steps. Reforms may lead to lower prices but at the cost of a large number of financially distressed firms and it is in this scenario that the government would be in a predicament. In such a situation it would help if one is able to identify a particular type of firm (defined by some common firm characteristic, size for example) that is less adversely affected than others, then the particular type of firms to be supported in the future is clear. For financial institutions too it would become extremely difficult to deal with a generalized financial distress amongst all types of firms compared to financial distress being concentrated in a particular part (i.e among a particular type of firms) of the industry.

To capture these aspects of the effects of reforms on firm performance we focus on firm level performance variables like market shares of the leading firms, profitability ratios (profit before interest depreciation and taxes as a ratio of sales) and export to sales ratio. To gauge the financial health of firms we look at variables like the interest coverage ratio (profit before interest and taxes but after depreciation divided by interest paid), ratio of external sources to total sources of funds, ratio of funds generated through the capital market to total external sources, and ratio of loans provided by Financial Institutions to total external funds. By looking at the sources and uses of funds allows us to trace how a firm raises resources.

We use two sources of data to study the effect of reforms. The first is the data from the annual survey of industries which we use to study the intermediate goods as a whole and the cement and steel industries at a broad industry level. The second is firm level data for cement and steel industries available through PROWESS, a data software provided by the Centre for Monitoring the Indian Economy (CMIE), Mumbai. The rest of the study is organized as follows. In Section II we provide a brief description of the policy framework surrounding the steel cement industries prior to 1990-1 and a brief description of the reform process. We describe the sectoral patterns for the intermediate goods as a whole and the steel and cement industries based on ASI data in Section III. Sections IV and V deal with the firm level picture of cement and steel industries respectively. Both these sections use data downloaded from PROWESS. Section VI is the conclusion.

## II. POLICY REGIMES

### *(A) Cement*

Price and distribution controls were first imposed on cement during World War II, when it was deemed an essential commodity. Controls, in one form or another, continued almost uninterruptedly until 1982. In 1956, the State Trading Corporation was invested with a monopoly right over the distribution of both domestically produced and imported cement. At the same time, a system of freight equalization was introduced, implying a uniform nationwide selling price of cement. In 1958, based on a Tariff Commission study, a three-tier retention price scheme was introduced, with producer prices ranging from Rs 54.50 to Rs 80.50 per ton, based on the usual considerations of vintage and technological conditions of the plants. The three-tier pricing scheme continued up to 1969, at which point retention prices ranged from Rs 90.50 to Rs 96.05 per ton. In 1966, the government lifted controls over the distribution of cement; the responsibility for distribution was collectively taken up by the cement manufacturers, who formed the Cement Allocation and Co-ordination Organization for this purpose. The government, however, reimposed distribution controls during the next year.

In 1969, the government replaced the three-tier scheme with a uniform retention price of Rs 100 per ton. This scheme continued until 1979, at which point the retention price was Rs 168.91 per ton, when a three-tier scheme was reintroduced. The prices paid to existing units ranged from Rs 185 to Rs 220 per ton; however, an added incentive was provided to new units and expanded capacity by way of a retention price of Rs 296 per ton. This was a relatively short-lived regime, lasting up to early 1982, when the government announced the partial removal of price and distribution controls. At this point, the retention prices for existing units ranged from Rs 233.39 to Rs 268.39 per ton. For expansions and new units, the price offered was Rs 344.39 per ton.

One of the major elements of the 1979 package was an emphasis on the setting up of so-called mini cement plants. These plants, basically scaled down versions of large plants, were considered appropriate for the exploitation of scattered limestone deposits which were too small to attract a large cement plant. They were further justified on the basis of their relatively low capital intensities, and, by virtue of their locations in backward areas, their ability to supply cement to their neighbouring areas, which again may not have been an attractive market for large cement plants located far away.

The reintroduction of the multi-tier formula in 1979, with its explicit incentive for new capacity, appears to have induced a surge of new investment in the cement industry, with installed capacity growing at 9 per cent per year, but most of this capacity had yet to begin producing by the time the price regime ended in 1982; production grew at only 2.7 per cent per year during this period.

In February 1982, the government introduced a system of partial decontrol of cement prices. Under this scheme, existing units were required to sell quantities of cement equalling 66.6 per cent of their installed capacity at a controlled or 'levy' price of Rs 335 per ton (of ordinary portland or portland slag cement). Any production above this levy amount could be sold at whatever price the market would bear. For sick units, the levy quota was set at 50 per cent of capacity, and for units commencing commercial production after 1 January 1982, the levy was set at 37.5 per cent in the first year of operation, 42.5 per cent in the second year and 50 per cent in subsequent years. In the years following partial decontrol, minor modifications were made to the existing structure, both by way of periodic increases in the levy price to compensate for various input price increases, and by way of small reductions in the levy quantity. In 1985, for instance, the basis of levy calculations was changed from the rated capacity of the unit to its actual production. The levy price regime was in place until 1989, when the government announced total decontrol of cement prices; all units were free to sell all their output at whatever price the market would bear.

*(B) Steel*

Historically speaking the Industrial Policy Resolution of 1948 had made it clear that the future of the steel industry was with the public sector. Investment in this sector took off in the second plan in which it was planned to triple the existing capacity. The actual commissioning of the first public sector steel plant took place at Bhilai in 1961, closely followed by the commissioning of the Rourkela and Durgapur plants in 1962. Construction of the now famous Bokaro steel plant started in 1966 and the plant was commissioned after much delay in 1972.

This industry was subject to all the forms of controls for example, reservation for public sector licensing, freight equalization, and price controls. The most complicated being freight equalization and price controls. Since independence the price was fixed by the tariff commission. In 1964 dual pricing system was introduced which differentiated between the prices of 'essential consumers' (defined as government, public sector enterprises and their suppliers and small scale industry) and other consumers. As compared to cement the price controls in steel are far more complicated because of the different types of steel involved. These prices were changed from time to time. An excellent description of the government policies with respect to steel can be found in Sidhu (1987) and Sengupta (1995).

In 1966 the government announced the delicensing of electric arc furnace steel units (or as known in common parlance as mini steel units). This was done because the integrated steel plants were not able to keep pace with the rise in demand for billets and demand for mild steel by the construction industry. The mini steel plants were also given an incentive in terms of excise duty rebates. This segment of the industry has grown quite rapidly over the years. In 1970 this segment was again brought under the licensing regime.

In 1972 the large public sector presence in steel was reorganized by the creation of the Steel Authority of India. In 1990-1 steel was delicensed and in 1992 it was freed from freight equalization and price controls. The integrated steel mills were automatically allowed to produce downstream products which was not possible prior to liberalization. In 1992-3 its import was freely allowed with a drastic reduction in import duty which in some categories exceeded 20 per cent. In the following years there was a further reduction in import duties. Nevertheless, two important inputs into this industry, coal (for the integrated steel plants) and electricity (for the mini steel plants) are still completely under government control.

## III. THE SECTORAL PATTERNS

We now turn to look at the broad sectoral picture. We begin by looking at the rates of growth of production of cement and steel (Fig. 1). In all the figures that follow we also report the relevant ratios for the intermediate goods sector as a whole. In the period

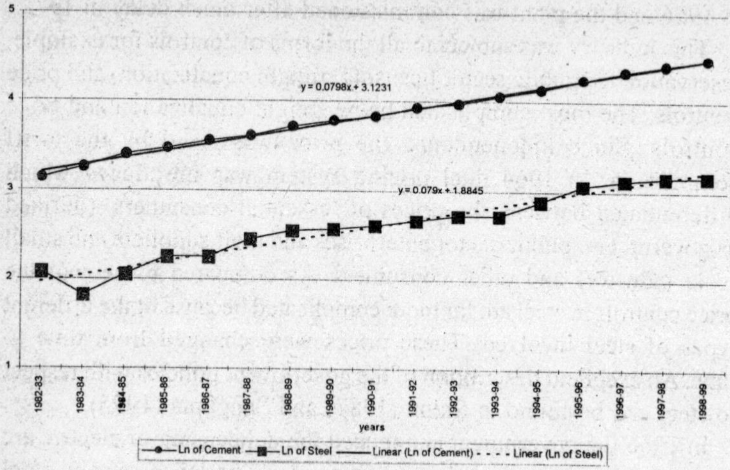


FIGURE 1. GROWTH RATE OF PRODUCTION

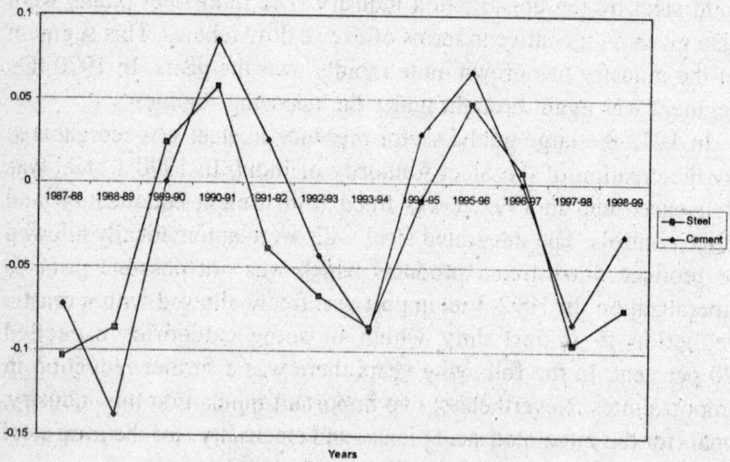


FIGURE 2. CHANGES IN REAL PRICE

1982-3 to 1997-8 the domestic production of both grew by about 8 per cent per annum. This can be considered to be a reasonably healthy rate of growth. The growth rate is estimated by fitting a exponential trend line. Very wild production swings have not been experienced and production has been steady. Both these industries are characterized by stability in production and a healthy growth rate of production in this period. Liberalization has not rendered the production of these commodities to become very volatile.

Next we examine the behaviour of the whole sale price index of both cement and steel. It should be noted here that cement prices were partially decontrolled in 1982 and full decontrol was effected in 1989. In the case of steel the decontrol of prices took effect from 1991. In Figure 2 we plot the rate of change in the wholesale price index of cement and steel over the previous year net of the rate of change the wholesale price index for manufactured goods. The real prices of both commodities spurted immediately after price decontrol. In the period after 1992-3 the real prices have risen only for two years in case of both the commodities. For the rest of the period the real prices have generally fallen. Overall it can be said that real prices have substantially fallen in the post-reform period. It needs to be recognized here that a fall in the real price of these commodities is unambiguously good from the point of view of the consumers of these commodities and this in some sense can be called a direct benefit of liberalization. If a majority of firms in these two industries were able to maintain their financial health in the face of these decreases in real prices (one possible way to achieve this would be through increases in productivity) then this would have no adverse implications for financial institutions and other providers of funds who might have lent money or invested in these sectors.

The reasons for this fall in real prices are likely to be different in the two cases. The fall in the real price of cement has been driven by large increases in capacity [for details see Gokarn and Vaidya (1993 and 1994)] and intense competition between firms. In the case of steel the fall in prices has been due to easing of restrictions on imports and lowering of import duties. Imports have steadily grown (though some yearly variation exists) after liberalization and in terms of market share they have constituted about 12 per cent to 14 per cent of the total market share for steel between 1992-3 and 1997-8. A healthy growth rate of domestic production seems to have contributed to the fall in real prices in both cases.

In Figures 3 and 4 we plot labour productivity (gross value added divided by mandays worked) and capital productivity (gross value added by capital employed) respectively. In both the cases, labour productivity has substantially increased in the post-reform period. Capital productivity on the other hand has been more or less stable

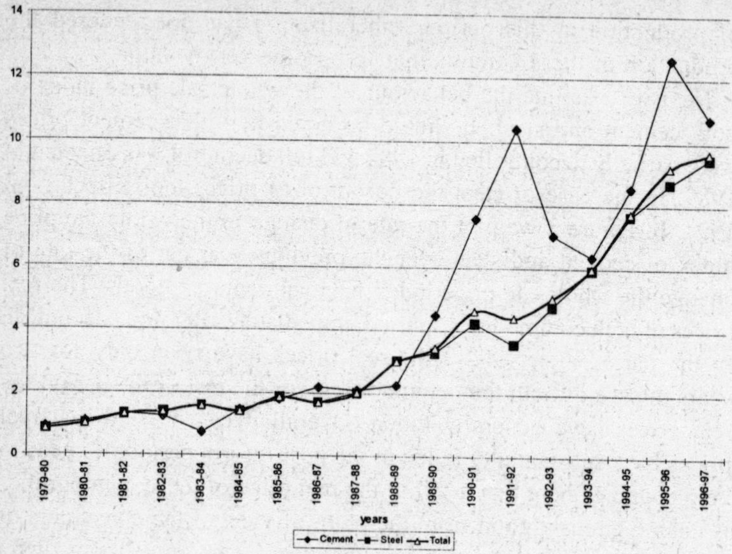


FIGURE 3. PRODUCTIVITY RATIOS
(GROSS VALUE ADDED/MANDAYS WORKED)

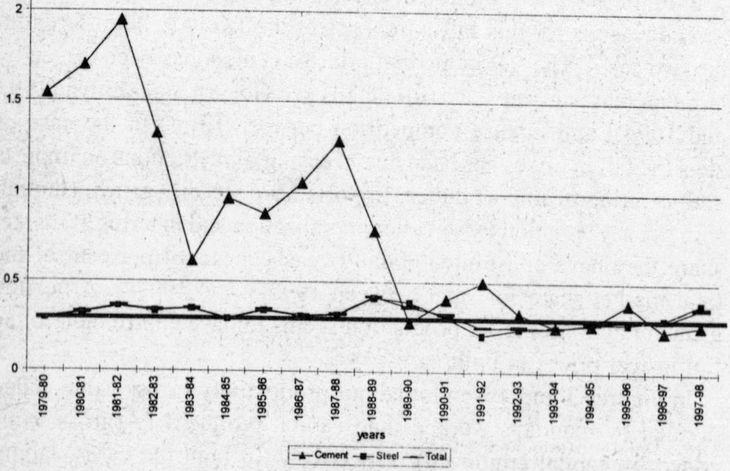


FIGURE 4. PRODUCTIVITY RATIOS
(GROSS VALUE ADDED/FIXED CAPITAL)

after liberalization though in the case of cement there were major increases in capital productivity prior to 1991. This was due to investment boom prior to 1991 combined with a major shift from the wet process (which was relatively less efficient) to the dry process [for details see Gokarn and Vaidya (1993 and 1994)]. We need to note here that fixed capital per plant has risen substantially in the post-liberalization period in both industries and this is what is likely to be driving the increases in labour productivity when capital productivity is stagnant. Real wages per worker are plotted in Figure 5 and this shows a major increase in real wages after 1992-3 in both cases though for cement the real wages are substantially lower than for steel. This may have something to do with the fact that a very large number of workers in the steel industry would be employed in the public sector enterprises and thus are highly unionized. Stagnant or declining real prices combined with rising real wages can obviously have a profound negative impact of financial health of firms unless accompanied by substantial increases in labour productivity. Though labour productivity has increased quite substantially in both the industries there are bound to be sharp inter-firm differences in the rise of labour productivity which would lead to substantial variation in inter-firm financial performance.

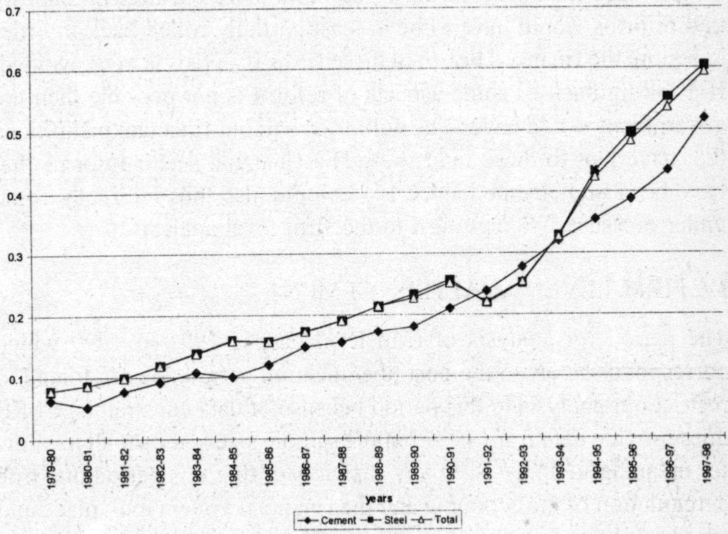


FIGURE 5. WAGES/WORKERS

At the broad sectoral level the indicators discussed above seem to suggest that reforms have generally had a positive impact especially on labour productivity, real wages and price levels. At this level of aggregation a lot of firm level details are lost. We have just noted that stagnant real prices and rising real wages are sure to put pressure on the financial health of firms. Moreover after liberalization both domestic competition and competition from imports have increased, further aggravating the situation. The reasons why there would be large inter-firm variations in financial health are many. In what follows we stress the role that size is likely to play in explaining inter-firm differences in financial performance but by no means is this the only or most important variable in all industries.

There are two scenarios that are likely to emerge, given that both steel and cement production are likely to involve large-scale economies. If large firms are more efficient (in the sense of having a lower per unit cost of production) then their financial health would be substantially better than relatively smaller firms. In such a case the loss making firms would be relatively small and on the whole this would not have a major impact on the institutions that lent money to them. If on the other hand, firms of all sizes are under financial stress after reforms then their implications for the financial sector as a whole are rather grim. In such a scenario there is a large probability that reforms would have to be at least partially rolled back in order to restore the financial health of these firms if a crisis is to be avoided. If a rolling back of some aspects of reform is not possible then the government would have to be willing to bail out financial institutions that have lent to these industries. The financial sector reforms that have been implemented since 1991 would also thus indirectly come under pressure. We now turn to the firm level analysis.

## IV. FIRM LEVEL ANALYSIS: CEMENT

The period for analysis of firm level data is 1991 to 1999 which corresponds solely to the post liberalization period. We are forced to restrict our analysis to this period because of data constraints. CMIE does provide data from 1989 but the sample size is too small to allow for meaningful analysis. In any case it is during this period that true deregulation of the economy and the financial system took place and thus looking at this period would enable us to ascertain the effects of liberalization on the financial health of firms.

TABLE 1. MARKET SHARES: CEMENT

| | *1992-3* | *1993-4* | *1994-5* | *1995-6* | *1996-7* | *1997-8* |
|---|---|---|---|---|---|---|
| Associated Cement Companies | 14.64 | 14.33 | 14.51 | 13.6 | 12.57 | 12.57 |
| Larsen and Toubro | 2.76 | 2.92 | 4.32 | 5.08 | 6.76 | 7.57 |
| Grasim Industries | 3.62 | 4.13 | 4.04 | 4.72 | 5.38 | 5.59 |
| Gujarat Ambuja Cements | 1.89 | 2.95 | 3.62 | 4.96 | 5.54 | 5.36 |
| Indian Rayon & Industries | 3.27 | 3.54 | 3.52 | 4.25 | 4.82 | 4.92 |
| Indian Cements | 4.71 | 4.87 | 4.95 | 4.62 | 4.42 | 4.81 |
| Century Textiles and Industries | 4.64 | 4.85 | 4.64 | 3.87 | 4.48 | 4.59 |
| Birla Corporation | 4.91 | 5.61 | 5.23 | 5.08 | 4.29 | 4.04 |
| Madras Cements | 2.23 | 2.36 | 2.42 | 2.58 | 2.45 | 2.79 |
| JK Synthetics | 3.91 | 3.87 | 3.84 | 3.81 | 3.17 | 2.71 |
| Cement Corporation of India | 6.01 | 5 | 3.54 | 2.98 | 2.65 | |
| Total for the above companies | 50.36 | 52.07 | 52.22 | 52.99 | 54.08 | 54.94 |
| Total for the sample of 116 companies | 95.56 | 95.39 | 99.86 | 97.65 | 94.92 | 96.06 |
| Imports | 0 | 0 | 0.01 | 0.01 | 0.01 | 0.01 |

SOURCE: *Markets and Market Shares*, Economic Intelligence Service, Bombay, Centre for Monitoring the Indian Economy, various issues.

We begin by looking at the market shares of the eleven largest firms over the period 1992-3 to 1997-8. The market shares are taken from the document *Markets and Market Shares* published by Centre for Monitoring the Indian Economy (CMIE), Mumbai for the relevant years. Unfortunately this data is not available after 1998. These are reported in Table 1. CMIE has a sample of 116 cement companies which account for above 96 per cent of total sale of cement in the country. The eleven firms we have chosen to concentrate on account for just over 50 per cent of total sales. It is clear that four firm concentration ratios (the conventional measure of concentration which is the sum of the market shares of the largest four firms) is not very high and is falling over the years. The ACC (Associated Cement Company) has always been the largest firm but its market share has steadily declined, while Larsen and Toubro, Gujarat Ambuja and Grasim Industries have substantially increased their market shares over the years. Clearly ACC's dominant position is under attack and it is only its head start (which has its roots in the Licensing Regime prior to 1990-1) is what is keeping it ahead. The turbulence in market shares indicates that competition for market shares across these firms is quite strong. This competitive activity can come down substantially if the industry starts experiencing a merger and acquisitions phase. If this happens we might even see rise in the real price of cement. With the watering down of the Monopolies and Restrictive Trade Practices (MRTP) act the ability of the government to influence mergers and acquisitions has drastically reduced. A competition policy is on the anvil but as of now there is no clear thinking on this issue on the part of the government.

We now turn to an analysis of the financial variables. The sample we work with varies in size from year to year. We take all those firms for which financial data was available after 1991. Thus for each year the number of firms in the sample varies. We compare these variables across large and small firms. A large firm is defined as one which has a market share of greater than one per cent. Using this definition 16 firms were classified as large up to 1995 and 15 firms thereafter. This is because of a merger that occurred in 1995. At this juncture it should be pointed out that this bifurcation into large and small is somewhat arbitrary but small changes in the this definition would not be expected to change our general results significantly. The variation in the number of firms in the sample is mainly in the sample of small firms.

In the tables that follow in the paper we present the number of firms in each (i.e. large and small) group, the minimum value, the maximum value, the average, the median and standard deviation in rows, two, three, four, five and six respectively. In the last three rows of the table we present a small frequency distribution of the variable. We follow this format of presenting the variable in all the tables that follow. Our intention is to get an idea of the entire distribution of the variable across firms.

In Tables 2a and 2b we present the ratio of sales from manufacturing to total sales. This variable gives an idea of how diversified the firm is into non-manufacturing activities. We see that in general small firms are more diversified though not to the extent so as to make a comparison amongst the financial variables suspect. It could well be that in certain cases being diversified into non-manufacturing activities would in fact be an advantage for the firm.

In Tables 3a and 3b we present the profitability ratio (PBDIT/sales, profit before depreciation interest and taxes) for small and large firms respectively. We use this particular definition of profitability because this ratio is immediately comparable across firms because it is not affected by the firms choice of leverage (as it is gross of interest) or tax payments and depreciation policies. This

TABLE 2a. RATIO OF MANUFACTURING SALES TO TOTAL SALES (CEMENT: LARGE FIRMS)

| | *1991* | *1992* | *1993* | *1994* | *1995* | *1996* | *1997* | *1998* | *1999* |
|---|---|---|---|---|---|---|---|---|---|
| Number of firms | 16.00 | 16.00 | 16.00 | 16.00 | 16.00 | 15.00 | 15.00 | 15.00 | 15.00 |
| Minimum | 0.27 | 0.54 | 0.57 | 0.56 | 0.60 | 0.84 | 0.66 | 0.40 | 0.39 |
| Maximum | 1.00 | 1.00 | 1.00 | 1.00 | 1.00 | 1.00 | 1.00 | 1.00 | 1.00 |
| Average | 0.95 | 0.96 | 0.96 | 0.94 | 0.94 | 0.96 | 0.94 | 0.94 | 0.93 |
| Median | 1.00 | 1.00 | 1.00 | 1.00 | 1.00 | 0.98 | 0.97 | 0.99 | 0.99 |
| Standard deviation | 0.18 | 0.11 | 0.11 | 0.11 | 0.11 | 0.06 | 0.09 | 0.15 | 0.15 |
| Zero | 6.25 | 6.25 | 6.25 | 6.25 | 6.25 | 0.00 | 6.67 | 6.67 | 6.67 |
| Greater than zero | 93.75 | 93.75 | 93.75 | 93.75 | 93.75 | 100.00 | 93.33 | 93.33 | 93.33 |

SOURCE: PROWESS, Centre for Monitoring the Indian Economy (CMIE), Mumbai.

TABLE 2b. RATIO OF MANUFACTURING SALES TO TOTAL SALES (CEMENT: SMALL FIRMS)

| | *1991* | *1992* | *1993* | *1994* | *1995* | *1996* | *1997* | *1998* | *1999* |
|---|---|---|---|---|---|---|---|---|---|
| Number of firms | 28.00 | 33.00 | 37.00 | 48.00 | 54.00 | 58.00 | 63.00 | 42.00 | 37.00 |
| Minimum | 0.00 | 0.00 | 0.00 | 0.00 | 0.00 | 0.00 | 0.00 | 0.00 | 0.00 |
| Maximum | 1.00 | 1.00 | 1.00 | 1.00 | 1.00 | 1.00 | 1.00 | 1.00 | 1.00 |
| Average | 0.95 | 0.93 | 0.89 | 0.83 | 0.84 | 0.87 | 0.85 | 0.91 | 0.87 |
| Median | 1.00 | 1.00 | 1.00 | 1.00 | 1.00 | 1.00 | 1.00 | 1.00 | 1.00 |
| Standard deviation | 0.19 | 0.18 | 0.25 | 0.34 | 0.32 | 0.30 | 0.33 | 0.26 | 0.31 |
| Zero | 3.57 | 3.03 | 8.11 | 18.75 | 18.52 | 12.07 | 14.29 | 9.52 | 10.81 |
| Greater than zero | 96.43 | 96.97 | 91.89 | 81.25 | 81.48 | 87.93 | 85.71 | 90.48 | 89.19 |

SOURCE: PROWESS, Centre for Monitoring the Indian Economy (CMIE), Mumbai.

TABLE 3a. PBDIT/SALES (CEMENT: LARGE FIRMS)

| | *1991* | *1992* | *1993* | *1994* | *1995* | *1996* | *1997* | *1998* | *1999* |
|---|---|---|---|---|---|---|---|---|---|
| Number of firms | 16.00 | 16.00 | 16.00 | 16.00 | 16.00 | 15.00 | 15.00 | 15.00 | 15.00 |
| Minimum | 0.06 | 0.17 | 0.05 | -0.10 | -0.10 | 0.03 | -0.08 | -0.02 | -0.26 |
| Maximum | 0.31 | 0.47 | 0.41 | 0.38 | 0.39 | 0.38 | 0.37 | 0.52 | 0.32 |
| Average | 0.21 | 0.25 | 0.19 | 0.16 | 0.20 | 0.20 | 0.17 | 0.18 | 0.14 |
| Median | 0.21 | 0.23 | 0.16 | 0.15 | 0.18 | 0.19 | 0.16 | 0.17 | 0.18 |
| Standard deviation | 0.07 | 0.09 | 0.10 | 0.11 | 0.12 | 0.09 | 0.12 | 0.14 | 0.15 |
| Zero | 0.00 | 0.00 | 0.00 | 6.25 | 6.25 | 0.00 | 6.67 | 6.67 | 13.33 |
| Greater than zero | 100.00 | 100.00 | 100.00 | 93.75 | 93.75 | 100.00 | 93.33 | 93.33 | 86.67 |

SOURCE: PROWESS, Centre for Monitoring the Indian Economy (CMIE), Mumbai.

ratio thus tells us how profitable the manufacturing activity has been in terms of recovering all direct costs (labour, raw-material and energy) except interest payments. If this ratio is positive then the firm does generate a surplus from its manufacturing activity but this may not in all cases be enough to cover interest payments, but a negative value

TABLE 3b. PBDIT/SALES (CEMENT: SMALL FIRMS)

| | *1991* | *1992* | *1993* | *1994* | *1995* | *1996* | *1997* | *1998* | *1999* |
|---|---|---|---|---|---|---|---|---|---|
| Number of firms | 28.00 | 33.00 | 37.00 | 48.00 | 54.00 | 58.00 | 63.00 | 42.00 | 37.00 |
| Minimum | -0.03 | -0.24 | -0.54 | -1.86 | -0.74 | -0.25 | -7.92 | -0.63 | -0.71 |
| Maximum | 0.33 | 4.86 | 0.57 | 0.49 | 3.99 | 4.43 | 1.35 | 0.27 | 1.23 |
| Average | 0.19 | 0.38 | 0.13 | 0.03 | 0.28 | 0.26 | -0.01 | 0.03 | 0.07 |
| Median | 0.21 | 0.23 | 0.18 | 0.11 | 0.19 | 0.19 | 0.10 | 0.07 | 0.11 |
| Standard deviation | 0.08 | 0.83 | 0.18 | 0.41 | 0.61 | 0.58 | 1.06 | 0.20 | 0.29 |
| Zero | 3.57 | 6.06 | 16.22 | 31.25 | 12.96 | 10.34 | 31.75 | 30.95 | 35.14 |
| Greater than zero | 96.43 | 93.94 | 83.78 | 68.75 | 87.04 | 89.66 | 68.25 | 69.05 | 64.86 |

SOURCE: PROWESS, Centre for Monitoring the Indian Economy (CMIE), Mumbai.

TABLE 4a. INTEREST COVERAGE RATIO (CEMENT: LARGE FIRMS)

| | *1991* | *1992* | *1993* | *1994* | *1995* | *1996* | *1997* | *1998* | *1999* |
|---|---|---|---|---|---|---|---|---|---|
| Number of firms | 16.00 | 16.00 | 16.00 | 16.00 | 16.00 | 15.00 | 15.00 | 15.00 | 15.00 |
| Minimum | -0.06 | 0.39 | 0.52 | -0.80 | -0.70 | -0.07 | -0.55 | -0.78 | -4.57 |
| Maximum | 4.90 | 6.57 | 5.31 | 3.80 | 10.06 | 13.07 | 7.32 | 3.82 | 2.27 |
| Average | 1.87 | 2.64 | 1.85 | 1.57 | 2.63 | 3.58 | 2.26 | 1.32 | 0.51 |
| Median | 1.63 | 1.96 | 1.46 | 1.24 | 2.10 | 3.06 | 2.08 | 1.31 | 1.08 |
| Standard deviation | 1.32 | 1.77 | 1.36 | 1.25 | 2.38 | 2.97 | 1.89 | 1.33 | 1.73 |
| Zero | 81.25 | 87.50 | 87.50 | 81.25 | 87.50 | 93.33 | 80.00 | 66.67 | 66.67 |
| Greater than zero | 18.75 | 12.50 | 12.50 | 18.75 | 12.50 | 6.67 | 20.00 | 33.33 | 33.33 |

SOURCE: PROWESS, Centre for Monitoring the Indian Economy (CMIE), Mumbai.

TABLE 4b. INTEREST COVERAGE RATIO (CEMENT: SMALL FIRMS)

| | *1991* | *1992* | *1993* | *1994* | *1995* | *1996* | *1997* | *1998* | *1999* |
|---|---|---|---|---|---|---|---|---|---|
| Number of firms | 28.00 | 33.00 | 37.00 | 48.00 | 54.00 | 59.00 | 64.00 | 42.00 | 38.00 |
| Minimum | -0.98 | -0.87 | -1.32 | -1.88 | -1.03 | -1.93 | -28.00 | -20.00 | -8.20 |
| Maximum | 22.35 | 17.51 | 20.83 | 16.80 | 19.58 | 41.00 | 17.62 | 21.57 | 10.37 |
| Average | 2.66 | 2.17 | 1.62 | 1.72 | 2.63 | 2.68 | 0.34 | 0.64 | 0.13 |
| Median | 0.95 | 1.11 | 1.14 | 0.80 | 1.47 | 1.92 | 0.28 | 0.10 | 0.07 |
| Standard deviation | 5.38 | 3.92 | 3.71 | 3.74 | 4.40 | 5.54 | 4.75 | 5.23 | 3.06 |
| Zero | 53.57 | 66.67 | 59.46 | 54.17 | 61.11 | 69.49 | 40.63 | 40.48 | 31.58 |
| Greater than zero | 46.43 | 33.33 | 40.54 | 45.83 | 38.89 | 30.51 | 59.38 | 59.52 | 68.42 |

SOURCE: PROWESS, Centre for Monitoring the Indian Economy (CMIE), Mumbai.

of this ratio indicates that the concerned firm is in financial distress because it is not even recovering its direct costs of production.

We see from the two tables that there is a distinct difference across the two groups of firms. Large firms in general had a distinctly higher profitability than small firms. The percentage of loss making firms is very large for small firms and this percentage has steadily increased over the period under consideration. For the large firms too this percentage increased sharply in 1999 but even then it was well below the percentage for small firms. The inter-firm variation (as shown by the standard deviation) is very high in the small firm group. Indeed there are some small firms that are extremely profitable. In general though the distinctly lower profitability of the small firms is bound to create problems for the financial institutions and banks from whom they have borrowed funds.

We next turn to the interest coverage ratio in Tables 4a and 4b. This is defined as PBIT (profits before interest and taxes but after depreciation) divided by interest payments. The idea here is to ask the following question: how many times can the firm make its interest payments out of the post-depreciation profits it earns. The higher this figure the healthier is the firm from the lender's perspective. A comparison between the two groups of firms clearly indicates that small firms have distinctly lower interest coverage ratio than large firms. The percentage of firms for whom this ratio is greater than 1 is significantly higher for large firms than small firms. In some sense this is only to be expected because small firms are less profitable. But this is not necessarily so because if small firms have a very low debt equity ratios a smaller profit may well cover the interest payments. What is extremely interesting is that the last three years in our sample period (1997 to 1999) are the years when the change in the real price of cement was either zero (1997) or negative (1998, 1999) and it is exactly for these years that the percentage of firms with the interest coverage ratio of less than one rose quite dramatically in both groups. The rise in this percentage was much larger for the small firms. This fall in the real price of cement could be explained by two factors. First, the general slow down in economic activity and slow down in investment in particular in this period and second, competitive rivalry among firms. As cement is a very thinly traded commodity and imports are negligible they had no role to play in this fall in real prices. Clearly, the fall in real prices have had a much larger adverse impact on smaller firms than larger firms.

As a measure of competitiveness of the firm's output we use the export to sales ratio. We report this ratio in Tables 5a and 5b. In general this ratio is low for the cement industry because of the very

TABLE 5a. EXPORT TO SALES RATIO (CEMENT: LARGE FIRMS)

| | *1991* | *1992* | *1993* | *1994* | *1995* | *1996* | *1997* | *1998* | *1999* |
|---|---|---|---|---|---|---|---|---|---|
| Number of firms | 16.00 | 15.00 | 15.00 | 17.00 | 16.00 | 16.00 | 16.00 | 15.00 | 15.00 |
| Minimum | 0.00 | 0.00 | 0.00 | 0.00 | 0.00 | 0.00 | 0.00 | 0.00 | 0.00 |
| Maximum | 0.07 | 0.06 | 0.09 | 0.11 | 1.00 | 0.14 | 0.17 | 0.17 | 0.11 |
| Average | 0.01 | 0.01 | 0.02 | 0.04 | 0.10 | 0.04 | 0.03 | 0.04 | 0.02 |
| Median | 0.00 | 0.00 | 0.00 | 0.02 | 0.02 | 0.01 | 0.01 | 0.01 | 0.01 |
| Standard deviation | 0.02 | 0.02 | 0.03 | 0.04 | 0.24 | 0.05 | 0.05 | 0.06 | 0.04 |
| Zero | 64.86 | 63.89 | 52.78 | 34.21 | 35.29 | 41.03 | 39.47 | 37.84 | 40.54 |
| Greater than zero | 35.14 | 36.11 | 47.22 | 65.79 | 64.71 | 58.97 | 60.53 | 62.16 | 59.46 |

*Source:* PROWESS, Centre for Monitoring the Indian Economy (CMIE), Mumbai.

TABLE 5b. EXPORT TO SALES RATIO (CEMENT: SMALL FIRMS)

| | *1991* | *1992* | *1993* | *1994* | *1995* | *1996* | *1997* | *1998* | *1999* |
|---|---|---|---|---|---|---|---|---|---|
| Number of firms | 27.00 | 32.00 | 36.00 | 47.00 | 54.00 | 58.00 | 61.00 | 42.00 | 35.00 |
| Minimum | 0.00 | 0.00 | 0.00 | 0.00 | 0.00 | 0.00 | 0.00 | 0.00 | 0.00 |
| Maximum | 0.02 | 0.01 | 0.05 | 1.00 | 1.00 | 1.00 | 1.00 | 1.00 | 1.00 |
| Average | 0.00 | 0.00 | 0.00 | 0.02 | 0.04 | 0.02 | 0.02 | 0.03 | 0.04 |
| Median | 0.00 | 0.00 | 0.00 | 0.00 | 0.00 | 0.00 | 0.00 | 0.00 | 0.00 |
| Standard deviation | 0.00 | 0.00 | 0.01 | 0.15 | 0.16 | 0.13 | 0.13 | 0.15 | 0.17 |
| Zero | 92.59 | 96.88 | 88.89 | 85.11 | 82.92 | 84.48 | 88.52 | 80.95 | 74.29 |
| Greater than zero | 7.41 | 3.13 | 11.11 | 14.89 | 17.08 | 15.52 | 11.48 | 19.05 | 25.71 |

SOURCE: PROWESS, Centre for Monitoring the Indian Economy (CMIE), Mumbai.

TABLE 6a. RATIO OF EXTERNAL SOURCES TO TOTAL SOURCES OF FUNDS (CEMENT: LARGE FIRMS)

| | *1991* | *1992* | *1993* | *1994* | *1995* | *1996* | *1997* | *1998* | *1999* |
|---|---|---|---|---|---|---|---|---|---|
| Number of firms | 16.00 | 15.00 | 15.00 | 16.00 | 16.00 | 16.00 | 16.00 | 15.00 | 15.00 |
| Minimum | -0.51 | -5.09 | -0.70 | -18.39 | -4.48 | -1.16 | -60.68 | -81.76 | -150.47 |
| Maximum | 0.82 | 0.71 | 1.67 | 0.91 | 17.19 | 14.42 | 3.06 | 1.12 | 117.19 |
| Average | -0.16 | -0.19 | 0.51 | -1.05 | 0.69 | 1.06 | -3.29 | -5.66 | -1.82 |
| Median | 0.08 | 0.16 | 0.50 | 0.31 | 0.46 | 0.43 | 0.62 | 0.54 | 0.49 |
| Standard deviation | 482.84 | 1.43 | 0.50 | 4.64 | 4.55 | 3.61 | 15.34 | 21.13 | 50.95 |
| Zero | 81.25 | 80.00 | 53.33 | 29.41 | 29.41 | 50.00 | 43.75 | 40.00 | 46.67 |
| Greater than zero | 18.75 | 20.00 | 46.67 | 70.59 | 70.59 | 50.00 | 56.25 | 60.00 | 53.33 |

SOURCE: PROWESS, Centre for Monitoring the Indian Economy (CMIE), Mumbai.

high transportation costs. Nevertheless on an average, the larger firms export more and a very large percentage of small firms do not export at all.

All the ratios considered above have been derived from the profit and loss accounts of firms. These ratios tell us something about the

TABLE 6b. RATIO OF EXTERNAL SOURCES TO TOTAL SOURCES OF FUNDS (CEMENT: SMALL FIRMS)

| | *1991* | *1992* | *1993* | *1994* | *1995* | *1996* | *1997* | *1998* | *1999* |
|---|---|---|---|---|---|---|---|---|---|
| Number of firms | 24.00 | 32.00 | 37.00 | 45.00 | 54.00 | 60.00 | 60.00 | 42.00 | 34.00 |
| Minimum | -0.58 | -4.27 | -5.34 | -9.91 | -29.00 | -6.50 | -38.00 | -38.33 | -136.00 |
| Maximum | 3.30 | 1.35 | 14.92 | 6.04 | 91.00 | 3.33 | 7.83 | 6.65 | 4.69 |
| Average | 0.64 | 0.31 | 0.96 | -0.28 | 3.69 | 0.31 | -1.02 | -0.81 | -4.86 |
| Median | 0.59 | 0.57 | 0.76 | 0.00 | 0.68 | 0.58 | 0.63 | 0.64 | -0.45 |
| Standard deviation | 0.85 | 0.98 | 2.72 | 2.79 | 17.51 | 1.51 | 7.60 | 6.57 | 23.40 |
| Zero | 25.00 | 31.25 | 21.62 | 55.56 | 53.24 | 33.33 | 31.67 | 28.57 | 54.55 |
| Greater than zero | 75.00 | 68.75 | 78.38 | 44.44 | 46.76 | 66.67 | 68.33 | 77.43 | 45.45 |

SOURCE: PROWESS, Centre for Monitoring the Indian Economy (CMIE), Mumbai.

TABLE 7a. RATIO OF FUNDS RAISED VIA CAPITAL MARKETS TO TOTAL EXTERNAL SOURCES (CEMENT: LARGE FIRMS)

| | *1991* | *1992* | *1993* | *1994* | *1995* | *1996* | *1997* | *1998* | *1999* |
|---|---|---|---|---|---|---|---|---|---|
| Number of firms | 16.00 | 15.00 | 15.00 | 17.00 | 16.00 | 16.00 | 16.00 | 15.00 | 15.00 |
| Minimum | -7.63 | -0.78 | -0.44 | -4.31 | -0.49 | -1.30 | -1.03 | -0.30 | -4.76 |
| Maximum | 0.53 | 3.11 | 0.97 | 3.26 | 1.71 | 1.31 | 1.61 | 3.75 | 4.86 |
| Average | -0.54 | 0.23 | 0.18 | 0.39 | 0.37 | 0.05 | 0.01 | 0.56 | 0.12 |
| Median | 0.02 | 0.01 | 0.00 | 0.07 | 0.25 | -0.01 | 0.00 | 0.18 | 0.00 |
| Standard deviation | 1.96 | 0.88 | 0.36 | 1.55 | 0.56 | 0.54 | 0.53 | 1.00 | 1.85 |
| Zero | 43.75 | 33.33 | 53.33 | 35.29 | 36.24 | 68.75 | 56.25 | 33.33 | 53.33 |
| Greater than zero | 56.25 | 66.67 | 46.67 | 64.71 | 63.76 | 31.25 | 43.75 | 66.67 | 46.67 |

SOURCE: PROWESS, Centre for Monitoring the Indian Economy (CMIE), Mumbai.

TABLE 7b. RATIO OF FUNDS RAISED VIA CAPITAL MARKETS TO TOTAL EXTERNAL SOURCES (CEMENT: SMALL FIRMS)

| | *1991* | *1992* | *1993* | *1994* | *1995* | *1996* | *1997* | *1998* | *1999* |
|---|---|---|---|---|---|---|---|---|---|
| Number of firms | 24.00 | 31.00 | 37.00 | 45.00 | 54.00 | 60.00 | 60.00 | 42.00 | 34.00 |
| Minimum | -0.21 | -0.61 | -0.21 | -1.60 | -3.14 | -3.06 | -6.75 | -2.29 | -4.03 |
| Maximum | 1.00 | 2.40 | 4.54 | 1.62 | 2.97 | 18.44 | 3.25 | 7.32 | 4.39 |
| Average | 0.08 | 0.16 | 0.28 | 0.06 | 0.18 | 0.85 | 0.04 | 0.28 | 0.14 |
| Median | 0.00 | 0.00 | 0.00 | 0.00 | 0.01 | 0.00 | 0.00 | 0.00 | 0.00 |
| Standard deviation | 0.22 | 0.48 | 0.78 | 0.42 | 0.83 | 3.15 | 1.11 | 1.24 | 1.18 |
| Zero | 62.5 | 61.29 | 51.35 | 71.11 | 62.5 | 53.33 | 58.33 | 64.29 | 70.59 |
| Greater than zero | 37.5 | 38.71 | 48.65 | 28.89 | 37.5 | 46.67 | 41.67 | 35.71 | 29.41 |

SOURCE: PROWESS, Centre for Monitoring the Indian Economy (CMIE), Mumbai.

financial health of firms. The financial health of firms is affected by the structure of its sources and uses of funds. We now turn to an analysis of the sources and uses of funds statements of the firms in the cement industry.

We consider in Tables 6a and 6b the ratio of external sources to total sources of funds. External sources consist of all sources of funds other than retained earnings and depreciation (which are internal sources). External sources involve some financial intermediary or financial market. The ability of firms to raise money through external sources many times reflects its financial position. On the other hand if a firm depends heavily on debt (rather new equity) then this may have a negative influence on its interest coverage ratio in following periods. Generally speaking the sources and uses of funds statements are notorious for extreme values. In the tables that follow this will be obvious, as the median becomes a more meaningful measure of central tendency. Even then interpretation is often difficult. The tables indicate that overall there does seem to be a significant difference in the extent to which external funds are tapped by both groups of firms though there exist significant year to year changes. The percentage of firms which do not use external sources at all is quite comparable across the two groups and this is only to be expected as many of the subcategories of external funds are by their nature tapped only infrequently. For example, new shares and loans from financial institutions are the two most important parts of external funds and these sources are used quite infrequently though the amounts raised by these sources are almost always very large. We turn to these two sources next.

First we look at the ratio of funds raised via capital markets to the total external sources in Tables 7a and 7b. The category capital markets includes issue of new shares, bonds and debentures and fixed deposits. This category is used much more by large firms than by small firms. The percentage of firms not tapping this source at all is very large for small firms. This is only to be expected because the types of information asymmetries that characterize these markets would make it very difficult for small firms to tap these markets, financial liberalization notwithstanding. In Tables 8a and 8b we look at the ratio of loans from financial institutions to total external sources. Here too, more or less the same picture emerges. Large firms seem to be at a significant advantage while taking loans from the financial institution. The percentage of firms not using this source at all is significantly larger for small firms.

The overall picture that emerges is that in the cement industry the large firms are at distinct advantage in the post-liberalization period.

TABLE 8a. RATIO OF LOANS FROM FIs TO TOTAL EXTERNAL SOURCES (CEMENT: LARGE FIRMS)

| | *1991* | *1992* | *1993* | *1994* | *1995* | *1996* | *1997* | *1998* | *1999* |
|---|---|---|---|---|---|---|---|---|---|
| Number of firms | 16.00 | 15.00 | 15.00 | 17.00 | 16.00 | 16.00 | 16.00 | 15.00 | 15.00 |
| Minimum | -1.43 | -1.52 | -0.74 | -1.23 | -2.86 | -0.96 | -0.39 | -0.63 | -19.92 |
| Maximum | 6.72 | 8.59 | 1.36 | 2.80 | 0.49 | 6.65 | 10.14 | 6.62 | 3.93 |
| Average | 0.47 | 0.63 | 0.09 | 0.18 | -0.26 | 0.76 | 1.65 | 0.82 | -0.99 |
| Median | 0.04 | 0.19 | 0.04 | 0.01 | 0.00 | 0.45 | 0.62 | 0.24 | 0.28 |
| Standard deviation | 1.77 | 2.31 | 0.49 | 0.88 | 0.90 | 1.71 | 3.23 | 1.83 | 5.40 |
| Zero | 37.50 | 33.33 | 46.67 | 47.06 | 37.50 | 31.25 | 12.50 | 33.33 | 33.33 |
| Greater than zero | 62.50 | 66.67 | 53.33 | 52.94 | 62.50 | 68.75 | 87.50 | 66.67 | 66.67 |

SOURCE: PROWESS, Centre for Monitoring the Indian Economy (CMIE), Mumbai.

TABLE 8b. RATIO OF LOANS FROM FIs TO TOTAL EXTERNAL SOURCES (CEMENT: SMALL FIRMS)

| | *1991* | *1992* | *1993* | *1994* | *1995* | *1996* | *1997* | *1998* | *1999* |
|---|---|---|---|---|---|---|---|---|---|
| Number of firms | 24.00 | 31.00 | 37.00 | 45.00 | 54.00 | 60.00 | 60.00 | 42.00 | 34.00 |
| Minimum | -0.37 | -12.60 | -3.63 | -7.91 | -1.66 | -42.00 | -3.57 | -3.71 | -23.27 |
| Maximum | 5.44 | 1.56 | 2.71 | 11.92 | 5.00 | 15.84 | 9.51 | 1.91 | 1.36 |
| Average | 0.90 | -0.05 | -0.01 | 0.36 | 0.18 | 0.38 | 0.36 | -0.05 | -0.96 |
| Median | 0.33 | 0.06 | 0.00 | 0.00 | 0.00 | 0.15 | 0.00 | 0.00 | 0.00 |
| Standard deviation | 1.46 | 2.38 | 0.96 | 2.25 | 1.08 | 6.37 | 1.81 | 0.91 | 4.23 |
| Zero | 29.17 | 48.39 | 56.76 | 66.67 | 54.76 | 41.67 | 63.33 | 66.67 | 67.65 |
| Greater than zero | 70.83 | 51.61 | 43.24 | 33.33 | 45.24 | 58.33 | 36.67 | 33.33 | 32.35 |

SOURCE: PROWESS, Centre for Monitoring the Indian Economy (CMIE), Mumbai.

The lowering of real prices and increase in competitive activity in this industry have led to a bias towards large firms. This is clearly an outcome of the inherent economies of scale in cement production. It is quite likely that all the increases in labour productivity have been captured by large firms which helped them to overcome the increase in real wages and decreases in real prices of cement. These advantages have manifested first in a higher profitability and greater overall financial health as compared to small firms, and second in a superior ability of large firms to tap external sources of funds compared to small firms. The financial ill health of small firms would very clearly have implications for financial institutions who have lent money to them. Luckily, the amounts involved are not large and the exposure of financial institution to the cement industry alone is not exceedingly large. The losses imposed on the financial institutions by the small firms can be absorbed by them and thus no serious externality is likely to arise due to this. The fact that weak firms are small could well be a major advantage for the cement industry and financial institutions who have lent to this industry.

## V. FIRM LEVEL ANALYSIS: STEEL

At the outset we need to recognize that the structure of the steel industry is far more complicated than the cement industry. Finished steel (this is the product we refer to in the study) can be produced in two ways. First it can be produced by melting the crude steel produced in the steel melting process (i.e. from integrated steel plants) and by re-rolling steel scrap (this is many times referred to as the mini steel sector. We have had on occasion pointed out that the technology involved in both these is fundamentally different. In terms of size the integrated steel plants are very much larger than the mini steel plants to the extent that any comparison between the two could be problematic. Our perspective here is that given the fact that for the consumer it is quite irrelevant where and how the steel is produced, these very different types of firms are ultimately competing with each other in the market. Moreover imports of finished steel would influence the performance of both the types of firms. In such a situation we need to study how liberalization has affected these firms.

We begin by looking at market shares of finished steel in Table 9. The market shares are taken from the document *Markets and Market Shares* published by Centre for Monitoring the Indian Economy (CMIE), Mumbai for the relevant years. The market share of Steel Authority of India Limited has declined quite substantially in this period. This is quite similar to what ACC experienced in the post-liberalization period. Tata Iron and Steel Company Limited on the other hand has maintained its market share over the period and the Rashtriya Ispat Nigam Limited has maintained its market share. These are the three firms with integrated steel plants. These three had a combined market share of about 62 per cent in 1992-3 which declined to just under 50 per cent in 1997-8. The re-rollers or mini steel plants as a group have gained market share. The market share data is based on a sample of 226 companies which combined with imports covers the entire market in the period under consideration. The turbulence in market shares suggests that competition has increased in this industry too.

TABLE 9. MARKET SHARES: STEEL

| | *1992-3* | *1993-4* | *1994-5* | *1995-6* | *1996-7* | *1997-8* |
|---|---|---|---|---|---|---|
| Steel Authority of India Ltd. | 47.49 | 47.08 | 39.7 | 36.29 | 32.66 | 31.52 |
| Tata Iron and Steel Co. | 10.66 | 10.81 | 10.75 | 10.68 | 10.57 | 9.83 |
| Rashtriya Ispat Nigam | 3.65 | 4.88 | 6.01 | 7.11 | 7.07 | 7.84 |
| Essar Steel | | | | 2.52 | 5.76 | 6.6 |
| Jindal Iron and Steel Co. | 0.57 | 0.95 | 1.92 | 2.06 | 2.18 | 2.68 |
| Jindal Strips | 2.52 | 2.38 | 2.29 | 2.34 | 1.95 | 2.29 |
| Lloyds Steel Industries | | | 0.82 | 1.9 | 2.5 | 2.04 |
| Ispat Industries | 1.94 | 1.92 | 2.02 | 1.77 | 1.84 | 1.92 |
| Usha Ispat | 0.82 | 0.71 | 0.58 | 0.49 | 0.82 | 1.24 |
| Uttam Steel | 0.82 | 0.68 | 1.11 | 1.06 | 1.13 | 1.18 |
| Total of the above companies | 68.45 | 69.41 | 65.2 | 66.23 | 66.48 | 67.13 |
| Total for the sample of 226 companies | 87.82 | 88.52 | 86.41 | 85.96 | 86.3 | 85.78 |
| Imports | 12.18 | 11.48 | 13.59 | 14.04 | 13.7 | 14.22 |

SOURCE: *Markets and Market Shares*, Economic Intelligence Service, Bombay, Centre for Monitoring the Indian Economy, various Issues.

The question of how to bifurcate this industry into two groups on the basis of size is not an easy one to tackle. One option would be to consider the three integrated steel plant firms as one group and club all the other firms into another group. While it is true that integrated steel plants would naturally be the larger group, such a bifurcation would club all the mini steel plants together irrespective of size and there are large mini steel plants too. As our main focus is on size and not on technology per se we decided to group the largest ten firms in 1999 together and classify them as large firms. The rest of the firms are considered to be small firms. As before one could criticize this as being arbitrary but it is unlikely that our results would substantially be altered with a slightly different grouping rule. Let us now turn to the financial variables.

We begin with the manufacturing sales to total sales ratio in Tables 10a and 10b. the ratios are quite similar for the two groups of firms indicating the diversification into non-manufacturing activities is not the prerogative of either type of firms. The fact that for some years the minimum values are zeros in this table indicates that for some years some firms did not produce any output at all. This is an extremely serious matter especially with respect to large firms indicating that some large firms are likely to be in major financial trouble. Such firms exist in both the groups. In both groups about 80 per cent of firms had a ratio between 0.8 and 1.

TABLE 10a. RATIO OF MANUFACTURING SALES TO TOTAL SALES (STEEL: LARGE FIRMS)

| | *1991* | *1992* | *1993* | *1994* | *1995* | *1996* | *1997* | *1998* | *1999* |
|---|---|---|---|---|---|---|---|---|---|
| Number of firms | 8 | 9 | 10 | 10 | 10 | 10 | 9 | 10 | 10 |
| Minimum | 0.00 | 0.60 | 0.00 | 0.00 | 0.23 | 0.45 | 0.44 | 0.28 | 0.18 |
| Maximum | 1.00 | 0.98 | 1.00 | 1.00 | 1.00 | 0.99 | 1.00 | 1.00 | 1.00 |
| Average | 0.81 | 0.89 | 0.81 | 0.79 | 0.80 | 0.83 | 0.84 | 0.84 | 0.83 |
| Median | 0.92 | 0.94 | 0.93 | 0.88 | 0.91 | 0.90 | 0.87 | 0.92 | 0.90 |
| Standard deviation | 0.33 | 0.12 | 0.30 | 0.30 | 0.25 | 0.18 | 0.17 | 0.22 | 0.25 |
| Zero | 12.50 | 11.11 | 30.00 | 40.00 | 30.00 | 30.00 | 22.22 | 20.00 | 20.00 |
| Greater than zero | 87.50 | 88.89 | 70.00 | 60.00 | 70.00 | 70.00 | 77.78 | 80.00 | 80.00 |

SOURCE: PROWESS, Centre for Monitoring the Indian Economy (CMIE), Mumbai.

TABLE 10b. RATIO OF MANUFACTURING SALES TO TOTAL SALES (CEMENT: SMALL FIRMS)

| | *1991* | *1992* | *1993* | *1994* | *1995* | *1996* | *1997* | *1998* | *1999* |
|---|---|---|---|---|---|---|---|---|---|
| Number of firms | 71 | 82 | 89 | 117 | 147 | 154 | 137 | 117 | 94 |
| Minimum | 0.00 | 0.00 | 0.00 | 0.00 | 0.00 | 0.00 | 0.00 | 0.00 | 0.00 |
| Maximum | 1.00 | 1.00 | 1.00 | 1.00 | 1.00 | 1.00 | 1.00 | 1.00 | 1.00 |
| Average | 0.91 | 0.85 | 0.86 | 0.79 | 0.83 | 0.84 | 0.86 | 0.81 | 0.83 |
| Median | 0.99 | 0.98 | 0.98 | 0.97 | 0.98 | 0.98 | 0.98 | 0.97 | 0.97 |
| Standard deviation | 0.23 | 0.29 | 0.29 | 0.34 | 0.31 | 0.29 | 0.25 | 0.30 | 0.29 |
| Zero | 7.04 | 14.63 | 13.48 | 23.08 | 17.69 | 14.94 | 17.52 | 23.93 | 20.21 |
| Greater than zero | 92.96 | 85.37 | 86.52 | 76.92 | 82.31 | 85.06 | 82.48 | 76.07 | 79.79 |

SOURCE: PROWESS, Centre for Monitoring the Indian Economy (CMIE), Mumbai.

TABLE 11a. PBDIT/SALES (STEEL: LARGE FIRMS)

| | *1991* | *1992* | *1993* | *1994* | *1995* | *1996* | *1997* | *1998* | *1999* |
|---|---|---|---|---|---|---|---|---|---|
| Number of firms | 8 | 9 | 10 | 10 | 10 | 10 | 9 | 10 | 10 |
| Minimum | 0.12 | -0.15 | 0.01 | 0.01 | 0.13 | 0.12 | 0.12 | 0.10 | 0.04 |
| Maximum | 0.70 | 0.42 | 0.48 | 0.41 | 0.72 | 0.32 | 0.32 | 0.33 | 0.21 |
| Average | 0.27 | 0.17 | 0.20 | 0.19 | 0.24 | 0.20 | 0.19 | 0.18 | 0.13 |
| Median | 0.15 | 0.17 | 0.17 | 0.17 | 0.18 | 0.20 | 0.18 | 0.16 | 0.14 |
| Standard Deviation | 0.22 | 0.15 | 0.15 | 0.13 | 0.18 | 0.07 | 0.06 | 0.07 | 0.06 |
| Zero | 0.00 | 11.11 | 0.00 | 0.00 | 0.00 | 0.00 | 0.00 | 0.00 | 0.00 |
| Greater than zero | 100.00 | 88.89 | 100.00 | 100.00 | 100.00 | 100.00 | 100.00 | 100.00 | 100.00 |

SOURCE: PROWESS, Centre for Monitoring the Indian Economy (CMIE), Mumbai.

In Tables 11a and 11b we present the PBDIT to sales ratios. The tables unambiguously point to the fact as a group the large firms have a substantially higher profitability than the small firms. In fact in recent years all the large firms had a PBDIT/sales ratio which was

TABLE 11b. PBDIT/SALES (STEEL: SMALL FIRMS)

| | 1991 | 1992 | 1993 | 1994 | 1995 | 1996 | 1997 | 1998 | 1999 |
|---|---|---|---|---|---|---|---|---|---|
| Number of firms | 71 | 82 | 89 | 117 | 147 | 154 | 137 | 117 | 94 |
| Minimum | -8.00 | -0.46 | -0.38 | -0.89 | -76.67 | -3.38 | -59.25 | -2.34 | -18.55 |
| Maximum | 0.23 | 0.23 | 42.67 | 1.39 | 15.50 | 4.13 | 3.85 | 2.82 | 1.00 |
| Average | -0.03 | 0.09 | 0.80 | 0.07 | -0.28 | 0.09 | -0.44 | 0.01 | -0.32 |
| Median | 0.08 | 0.09 | 0.08 | 0.06 | 0.09 | 0.08 | 0.06 | 0.05 | 0.06 |
| Standard deviation | 0.96 | 0.09 | 4.73 | 0.21 | 6.49 | 0.50 | 5.10 | 0.43 | 2.15 |
| Zero | 9.86 | 10.98 | 12.36 | 23.93 | 12.24 | 16.88 | 32.85 | 27.35 | 28.72 |
| Greater than zero | 90.14 | 89.02 | 87.64 | 76.07 | 87.76 | 83.12 | 67.15 | 72.65 | 71.28 |

SOURCE: PROWESS, Centre for Monitoring the Indian Economy (CMIE), Mumbai.

TABLE 12a. INTEREST COVERAGE RATIO (STEEL: LARGE FIRMS)

| | 1991 | 1992 | 1993 | 1994 | 1995 | 1996 | 1997 | 1998 | 1999 |
|---|---|---|---|---|---|---|---|---|---|
| Number of firms | 8 | 9 | 10 | 10 | 10 | 10 | 9 | 10 | 10 |
| Minimum | 0.00 | -1.06 | -1.33 | -0.57 | 0.05 | 0.35 | 1.01 | -0.04 | -1.16 |
| Maximum | 3.06 | 3.47 | 5.27 | 5.98 | 5.82 | 3.76 | 4.14 | 2.81 | 1.82 |
| Average | 1.93 | 1.71 | 2.00 | 2.49 | 2.70 | 2.41 | 2.02 | 1.31 | 0.61 |
| Median | 1.95 | 1.71 | 1.74 | 1.95 | 2.59 | 2.63 | 1.59 | 1.17 | 0.93 |
| Standard deviation | 0.97 | 1.21 | 1.94 | 2.11 | 1.46 | 0.98 | 1.01 | 0.81 | 0.87 |
| Zero | 87.50 | 88.89 | 80.00 | 80.00 | 90.00 | 90.00 | 100.00 | 80.00 | 60.00 |
| Greater than zero | 12.50 | 11.11 | 20.00 | 20.00 | 10.00 | 10.00 | 0.00 | 20.00 | 40.00 |

SOURCE: PROWESS, Centre for Monitoring the Indian Economy (CMIE), Mumbai.

TABLE 12b. INTEREST COVERAGE RATIO (STEEL: SMALL FIRMS)

| | 1991 | 1992 | 1993 | 1994 | 1995 | 1996 | 1997 | 1998 | 1999 |
|---|---|---|---|---|---|---|---|---|---|
| Number of firms | 71 | 82 | 89 | 117 | 146 | 154 | 135 | 115 | 93 |
| Minimum | -10.72 | -1.91 | -9.52 | -6.58 | -13.00 | -496.00 | -650.00 | -169.50 | -71.14 |
| Maximum | 29.50 | 11.13 | 109.25 | 35.67 | 219.00 | 313.00 | 19.21 | 9.12 | 9.93 |
| Average | 1.78 | 1.69 | 2.68 | 1.78 | 3.54 | 0.30 | -3.85 | -1.44 | -0.88 |
| Median | 1.45 | 1.32 | 1.25 | 1.36 | 1.47 | 1.33 | 1.07 | 0.63 | 0.58 |
| Standard deviation | 3.92 | 1.96 | 11.79 | 4.16 | 18.36 | 47.55 | 56.15 | 17.03 | 9.75 |
| Zero | 77.46 | 73.17 | 60.67 | 61.54 | 65.75 | 62.34 | 55.56 | 46.09 | 44.09 |
| Greater than zero | 22.54 | 26.83 | 39.33 | 38.46 | 34.25 | 37.66 | 44.44 | 53.91 | 55.91 |

SOURCE: PROWESS, Centre for Monitoring the Indian Economy (CMIE), Mumbai.

positive. But this may not be as significant as it may first sound because what really matters is whether these firms are able to meet their interest payments. What the positive PBDIT to sales ratios do certainly indicate is that the larger firms are certainly reaping the benefits of scale economies. For the small firms quite a large percentage had zero or negative profits and the lack of scale economies would certainly have been a factor that contributed to this.

We present the interest coverage ratios in Tables 12a and 12b. The measures of central tendency generally indicate that this ratio is significantly lower for the smaller firms. The percentage of firms who have interest coverage ratio of less than one (i.e. firms unable to meet interest payments) is consistently larger for small firms. This percentage was very small among the large firms till 1997 after which it registered a sharp rise. It is pertinent to note that it is exactly for these years (1997-8 and 1998-9) that there was a sharp drop in the real price of steel. This sharp drop in the real price of steel, as in the case of cement was due to two factors. First the slow down in economic activity (this factor is similar in the case of cement) and second the sharp fall in the international prices of steel coupled with larger imports in these years have caused this drop in real prices. In some sense after the customs duty on imports of steel were reduced a rise in imports was inevitable.

Clearly like in the case of cement, the steel industry too was adversely affected by the drop in real prices and the smaller firms were affected much more. We need to note here that even if a few large firms are adversely affected this could have serious implication because of the sheer size of the large firms. If the largest firm, i.e. the Steel Authority of India Limited which produces over 30 per cent of finished steel gets into financial trouble (and it has) the implications of this are likely to be far more serious than some small firms getting into financial trouble. The sheer size of this firm, which to some extent driven by scale economies and to some extent by the choice made the government of India to create a monolith like SAIL, would be a disadvantage. So even though the percentage of firms that could not meet interest payments is higher for small firms the fact that some very large firms could not meet their interest payments is an extremely serious issue with possibly much larger overall financial implications than all the small firms put together. This fact about the steel industry stands in sharp contrast to the cement industry.

We turn to the export to sales ratios in Tables 13a and 13b. The average export to sales ratios are much smaller for large firms than small firms though the percentage of firms having no exports at all is much lower for large firms. Given the depressed price of steel in international markets these exports might have been driven more by

TABLE 13a. EXPORT TO SALES RATIO (STEEL: LARGE FIRMS)

| | *1991* | *1992* | *1993* | *1994* | *1995* | *1996* | *1997* | *1998* | *1999* |
|---|---|---|---|---|---|---|---|---|---|
| Number of firms | 7 | 9 | 10 | 10 | 10 | 10 | 9 | 10 | 10 |
| Minimum | 0 | 0 | 0 | 0 | 0 | 0 | 0 | 0 | 0 |
| Maximum | 0.1 | 0.17 | 0.33 | 0.44 | 0.73 | 0.7 | 0.04 | 0.05 | 0.04 |
| Average | 0.02 | 0.02 | 0.09 | 0.15 | 0.17 | 0.02 | 0.01 | 0.01 | 0.01 |
| Median | 0 | 0 | 0.02 | 0.065 | 0.085 | 0.07 | 0.05 | 0.11 | 0.09 |
| Standard deviation | 0.04 | 0.06 | 0.12 | 0.17 | 0.23 | 0.02 | 0.01 | 0.01 | 0.01 |
| Zero | 12.50 | 0.00 | 40.00 | 10.00 | 40.00 | 10.00 | 22.22 | 20.00 | 30.00 |
| Greater than zero | 87.50 | 100.00 | 60.00 | 90.00 | 60.00 | 90.00 | 77.78 | 80.00 | 70.00 |

SOURCE: PROWESS, Centre for Monitoring the Indian Economy (CMIE), Mumbai.

TABLE 13b. EXPORT TO SALES RATIO (STEEL: SMALL FIRMS)

| | *1991* | *1992* | *1993* | *1994* | *1995* | *1996* | *1997* | *1998* | *1999* |
|---|---|---|---|---|---|---|---|---|---|
| Number of firms | 71 | 82 | 89 | 116 | 147 | 154 | 136 | 115 | 93 |
| Minimum | 0.00 | 0.00 | 0.00 | 0.00 | 0.00 | 0.00 | 0.00 | 0.00 | 0.00 |
| Maximum | 0.22 | 1.00 | 1.00 | 1.00 | 0.78 | 1.83 | 1.39 | 1.43 | 1.37 |
| Average | 0.01 | 0.02 | 0.03 | 0.03 | 0.03 | 0.05 | 0.05 | 0.05 | 0.07 |
| Median | 0.00 | 0.00 | 0.00 | 0.00 | 0.00 | 0.00 | 0.00 | 0.00 | 0.00 |
| Standard deviation | 0.03 | 0.12 | 0.12 | 0.11 | 0.11 | 0.18 | 0.16 | 0.18 | 0.20 |
| Zero | 87.32 | 86.59 | 88.76 | 79.31 | 77.55 | 74.68 | 72.79 | 71.30 | 65.59 |
| Greater than zero | 12.68 | 13.41 | 11.24 | 20.69 | 22.45 | 25.32 | 27.21 | 28.70 | 34.41 |

SOURCE: PROWESS, Centre for Monitoring the Indian Economy (CMIE), Mumbai.

fiscal incentives given to exporters than any plant level efficiency advantages. This is not to say that all plants cannot produce steel competitive in international markets but that competitiveness in international markets is unlikely to be the driving force behind expert performance of a majority of steel producers.

We now turn to an analysis of the sources of funds statements of the firms in this industry. In Tables 14a and 14b we present the ratio of external sources to total funds raised. Here again the presence of extreme values is obvious and we thus concentrate on the median. The large firms in general have been able to raise quite a large part of the total funds from external sources. In contrast small firms have relied more on internal sources as is obvious from the fact a substantial proportion of small firms raised no funds through external sources at all. There seems to be a general reluctance on the part of external providers of funds to provide resources to small firms. This tells us lot about the financial health of these firms in general and about the investment opportunities surrounding these firms in particular. What

TABLE 14a. RATIO OF EXTERNAL SOURCES TO TOTAL SOURCES OF FUNDS (STEEL: LARGE FIRMS)

| | *1991* | *1992* | *1993* | *1994* | *1995* | *1996* | *1997* | *1998* | *1999* |
|---|---|---|---|---|---|---|---|---|---|
| Number of firms | 8 | 9 | 10 | 10 | 10 | 10 | 9 | 10 | 10 |
| Minimum | 0.4 | 0.73 | 0 | 0.72 | 0.47 | 0.04 | 0.13 | -3.02 | 0.02 |
| Maximum | 1.29 | 1.49 | 1.57 | 5.49 | 0.95 | 2.25 | 0.08 | 1.12 | 2.86 |
| Average | 0.81 | 0.88 | 0.82 | 1.31 | 0.78 | 0.08 | 0.05 | 0.04 | 1.11 |
| Median | 0.81 | 0.81 | 0.84 | 0.845 | 0.825 | 0.755 | 0.53 | 0.765 | 0.9 |
| Standard deviation | 0.27 | 0.24 | 0.37 | 1.47 | 0.16 | 0.05 | 0.03 | 1.22 | 0.05 |
| Zero | 0.00 | 0.00 | 10.00 | 0.00 | 0.00 | 0.00 | 0.00 | 10.00 | 0.00 |
| Greater than zero | 100.00 | 100.00 | 90.00 | 100.00 | 100.00 | 100.00 | 100.00 | 90.00 | 100.00 |

SOURCE: PROWESS, Centre for Monitoring the Indian Economy (CMIE), Mumbai.

TABLE 14b. RATIO OF EXTERNAL SOURCES TO TOTAL SOURCES OF FUNDS (STEEL: SMALL FIRMS)

| | *1991* | *1992* | *1993* | *1994* | *1995* | *1996* | *1997* | *1998* | *1999* |
|---|---|---|---|---|---|---|---|---|---|
| Number of firms | 66 | 78 | 88 | 116 | 147 | 153 | 139 | 119 | 93 |
| Minimum | -7.57 | -8.28 | -20.75 | -35.25 | -8.95 | -3.00E +14 | -163.5 | -179.22 | -2.81E +14 |
| Maximum | 4.05 | 10.33 | 20.19 | 23.2 | 11.33 | 286 | 5.63E +13 | 2.59E +14 | 5.42E +14 |
| Average | 0.46 | 0.78 | 1.06 | 0.11 | 1.01 | -1.96E +12 | 4.05E +11 | 2.18E +12 | 2.80E +12 |
| Median | 0.715 | 0.765 | 0.83 | 0.725 | 0.83 | 0.75 | 0.61 | 0.66 | 0.78 |
| Standard deviation | 1.26 | 1.99 | 4.16 | 5.07 | 1.95 | 2.43E +13 | 4.77E +12 | 2.38E +13 | 6.36E +13 |
| Zero | 87.32 | 86.59 | 88.76 | 79.31 | 77.55 | 74.68 | 72.79 | 71.30 | 65.59 |
| Greater than zero | 12.68 | 13.41 | 11.24 | 20.69 | 22.45 | 25.32 | 27.21 | 28.70 | 34.41 |

SOURCE: PROWESS, Centre for Monitoring the Indian Economy (CMIE), Mumbai.

is worrying is that these firms' future prospects are bleak if there is such reluctance to provide funds to these firms. Moreover the internal funds available to these firms is going to very small compared to their needs, given their low profitability and size. Overall prospects of further investments by these firms seem pretty bleak and thus increases in productivity are unlikely to be realized.

Capital markets are an important source of external funds. In Tables 15a and 15b we present the ratio of funds raised through the capital market (consisting of new issues, bonds, debentures and fixed

TABLE 15a. RATIO OF FUNDS RAISED VIA CAPITAL MARKETS TO TOTAL EXTERNAL SOURCES (STEEL: LARGE FIRMS)

| | *1991* | *1992* | *1993* | *1994* | *1995* | *1996* | *1997* | *1998* | *1999* |
|---|---|---|---|---|---|---|---|---|---|
| Number of firms | 8 | 9 | 10 | 10 | 10 | 10 | 9 | 10 | 10 |
| Minimum | 0.02 | -0.01 | -1.31 | 0.27 | 0 | -0.01 | -2.17 | -6.72 | -0.002 |
| Maximum | 0.92 | 0.83 | 1.34 | 10.27 | 0.65 | 0.7 | 2.90 | 0.05 | 4.82E |
| Average | 0.31 | 0.36 | 0.27 | 1.71 | 0.28 | 0.02 | 0.05 | -0.05 | 0.08 |
| Median | 0.175 | 0.32 | 0.3 | 0.815 | 0.3 | 0.115 | 0.61 | -0.025 | 0.325 |
| Standard deviation | 0.32 | 0.32 | 0.72 | 3.03 | 0.21 | 0.02 | 1.34 | 2.16 | 1.46 |
| Zero | 0.00 | 22.22 | 40.00 | 0.00 | 20.00 | 40.00 | 11.11 | 70.00 | 20.00 |
| Greater than zero | 100.00 | 77.78 | 60.00 | 100.00 | 80.00 | 60.00 | 88.89 | 30.00 | 80.00 |

SOURCE: PROWESS, Centre for Monitoring the Indian Economy (CMIE), Mumbai.

TABLE 15b. RATIO OF FUNDS RAISED VIA CAPITAL MARKETS TO TOTAL EXTERNAL SOURCES (STEEL: SMALL FIRMS)

| | *1991* | *1992* | *1993* | *1994* | *1995* | *1996* | *1997* | *1998* | *1999* |
|---|---|---|---|---|---|---|---|---|---|
| Number of firms | 66 | 78 | 88 | 116 | 147 | 153 | 139 | 119 | 93 |
| Minimum | -0.13 | -15.58 | -3.72 | -7.13 | -1.88 | -2.49 | -5.89 | -1.52 | -6.84 |
| Maximum | 1.44 | 2.15 | 1.84 | 25 | 18.17 | 9.17 | 6.4 | 4.5 | 0.97 |
| Average | 0.10 | -0.14 | 0.09 | 0.37 | 0.35 | 0.22 | 0.13 | 0.09 | -0.05 |
| Median | 0 | 0 | 0 | 0 | 0 | 0 | 0 | 0 | 0 |
| Standard deviation | 0.27 | 1.80 | 0.70 | 2.45 | 1.54 | 0.94 | 0.96 | 0.47 | 0.74 |
| Zero | 87.32 | 86.59 | 88.76 | 79.31 | 77.55 | 74.68 | 72.79 | 71.30 | 65.59 |
| Greater than zero | 12.68 | 13.41 | 11.24 | 20.69 | 22.45 | 25.32 | 27.21 | 28.70 | 34.41 |

SOURCE: PROWESS, Centre for Monitoring the Indian Economy (CMIE), Mumbai.

TABLE 16a. RATIO OF LOANS FROM FIs TO TOTAL EXTERNAL SOURCES (STEEL: LARGE FIRMS)

| | *1991* | *1992* | *1993* | *1994* | *1995* | *1996* | *1997* | *1998* | *1999* |
|---|---|---|---|---|---|---|---|---|---|
| Number of firms | 8 | 9 | 10 | 10 | 10 | 10 | 9 | 10 | 10 |
| Minimum | -0.12 | 0 | -0.03 | -0.16 | -0.2 | -1.58 | -3.15 | -0.32 | -2.15 |
| Maximum | 0.54 | 1.02 | 0.54 | 2.08 | 0.38 | 0.25 | 1.43 | 0.07 | 14.3 |
| Average | 0.17 | 0.29 | 0.21 | 0.22 | 0.12 | -0.01 | -0.02 | 0.02 | 1.50 |
| Median | 0.155 | 0.2 | 0.23 | -0.065 | 0.13 | -0.01 | -0.07 | 0.095 | 0.08 |
| Standard deviation | 0.21 | 0.30 | 0.19 | 0.68 | 0.17 | 0.05 | 1.22 | 0.03 | 4.62 |
| Zero | 37.50 | 11.11 | 30.00 | 60.00 | 30.00 | 50.00 | 55.56 | 30.00 | 50.00 |
| Greater than zero | 62.50 | 88.89 | 70.00 | 40.00 | 70.00 | 50.00 | 44.44 | 70.00 | 50.00 |

SOURCE: PROWESS, Centre for Monitoring the Indian Economy (CMIE), Mumbai.

deposits) to total funds raised from external sources. Here too the advantage that larger firms have is obvious. Not only is this ratio in general higher for large firms but an extremely small percentage of

TABLE 16b. RATIO OF LOANS FROM FIs TO TOTAL EXTERNAL SOURCES (STEEL: SMALL FIRMS)

| | *1991* | *1992* | *1993* | *1994* | *1995* | *1996* | *1997* | *1998* | *1999* |
|---|---|---|---|---|---|---|---|---|---|
| Number of firms | 66 | 78 | 88 | 116 | 147 | 153 | 139 | 119 | 93 |
| Minimum | -0.49 | -5.85 | -3.39 | -1.17 | -1.96 | -3 | -8 | -4.1 | -0.84 |
| Maximum | 1.57 | 5.5 | 1.19 | 4.16 | 4.38 | 58.33 | 4.89 | 3.7 | 35 |
| Average | 0.17 | 0.13 | 0.06 | 0.18 | 0.06 | 0.57 | 0.01 | 0.11 | 0.74 |
| Median | 0 | 0 | 0 | 0 | 0 | 0 | 0 | 0 | 0 |
| Standard deviation | 0.38 | 1.01 | 0.54 | 0.57 | 0.53 | 4.91 | 1.10 | 0.69 | 3.86 |
| Zero | 53.03 | 58.97 | 65.91 | 59.48 | 68.03 | 62.75 | 61.15 | 64.71 | 50.54 |
| Greater than zero | 46.97 | 41.03 | 34.09 | 40.52 | 31.97 | 37.25 | 38.85 | 35.29 | 49.46 |

SOURCE: PROWESS, Centre for Monitoring the Indian Economy (CMIE), Mumbai.

small firms are able to raise funds from this source. The capital markets seem extremely biased towards large firms.

The ratio of funds raised from financial institutions to total external sources is presented in Tables 16a and 16b. The scenario here is slightly different than in the earlier case. In the case of large firms the amounts raised from this source after 1996 have fallen quite dramatically though prior to 1996 these firms have been able to raise funds from financial institutions. The simple averages are higher for small firms. Moreover about 40 per cent of small firms have been able to get loans from financial institutions in recent years. This may indicate that financial institutions have been asked by the government to lend to these firms. Given the general financial health of small firms (and also some large firms) a large portion of these loans are likely to turn into bad and doubtful debts. Herein lies the danger for the financial institutions.

Overall the picture for the steel industry is different from that of cement in one important aspect. There exist some very large firms who are not able to meet their interest payments and given the extreme asymmetry of size between large and small firms this can be very dangerous for the industry as a whole and the financial institutions who have lent them funds. It is in this sense that the fall in the real prices of steel and rise in real wages have not in all cases been mitigated by a rise in productivity both for large and small firms. In the post-reform period both large and small firms have been adversely affected though small firms have been hit harder. Finaneial distress is more evenly distributed across both the groups of firms. The advantages of scale economies have been enough to insulate the large firms from negative price shocks. It is quite possible that these price shocks have been very large (they are certainly larger than those experienced by cement industry, see Fig. 2) and this is the reason why its effects may be more evenly distributed across the industry. Whatever the reason, this has serious implications for the financial institutions who have lent funds to these firms. As pointed out in the introduction both IDBI and ICICI could be left with balance sheets badly mauled by the weak performance of steel firms.

## VI. CONCLUSION

In our attempt to access the effect of reforms on the cement and steel industries has yielded at the broad sectoral level the following important conclusions:

(a) In the post-reform period domestic production of both grew by about 8 per cent per annum. This can be considered a healthy growth rate by past standards.
(b) In the period after 1992-3 the real prices have risen only for two years in case of both the commodities. This decrease has been sharper for steel than for cement. This signifies that consumers have greatly benefited after liberalization.
(c) In both the cases labour productivity has substantially increased in the post-reform period. Capital productivity on the other hand has been more or less stable after liberalization. The gains in labour productivity are higher for cement than steel. There are again clear positive signs on the labor productivity front.
(d) After 1992-3 there was major increase in real wages in both cases though for cement the real wages are substantially lower than for steel.

Though all the above conclusions have a direct relevance for the financial health of firms in these two industries, (b) and (d) are of particular relevance. Notice that the steel industry faces a larger decrease in the real price combined with a higher real wage and a lower labour productivity as compared to the steel industry. This itself would suggest that the steel industry is very likely to have financial distress among firms that is far more evenly distributed across size classes. At the firm level this is exactly what we found.

More specifically at the firm level our major conclusions are:

(a) Across both the industries market shares of the leading firms (and concentration ratios) have fallen quite dramatically. This is certainly a benefit as generally speaking low concentration ratios imply lower monopoly power for the leading firms. This is extremely important in a liberalized set-up. With the substantial modification of the MRTP act, the commission has lost a lot of powers which enabled it to deal with issues arising out of industrial concentration in the past. With concentration level falling almost across the board after reforms this curtailment of powers of the commission is unlikely to be accompanied by abuse of monopoly power at least in the steel industry for certain. Moreover the international prices of steel act a strict disciplining device through imports. There is some scope for abuse of monopoly power in the cement industry even in the face of falling concentration ratios because aggregate market shares mask regional monopoly power which may arise because of high transportation costs. In this case the commission needs to be watchful and take action as required.

(b) Firms in the cement industry seem to be more diversified into non manufacturing activities than firms in the steel industry. And in both cases it is the small firms that are more diversified.

(c) In terms of profitability large firms across both the industries seem to be at a distinct advantage clearly reflecting scale economies in both industries. The difference is much more marked in the steel industry than in the cement industry.

(d) The advantage large firms have in terms of the interest coverage ratio suggests that financial distress is more evenly distributed across size classes of firms in the steel industry than in the cement industry.

(e) On the export front too it is the large firms that perform better. In the case of steel though average export to sales ratios are higher for small firms the percentage of small firms not exporting at all is very large.

(f) The analysis of the sources of funds statements suggests that larger firms have easier access to external funds in general. The smaller firms find it particularly difficult to access funds from capital markets. Loans from financial institutions seem easier to get for small firms though vis-à-vis the large firms they are at disadvantage with respect to both the sources. This calls into question the ability of small firms to make productivity enhancing investments in the future.

What is really serious is that if small firms in both the industries are unable to make investments required to increase productivity, their very survival in the medium term comes into question. As exit policies (specifically the bankruptcy procedure) have not been touched by the reform process this can cause further damage to financial institutions who lent to these firms. This is mainly because, as pointed out by Anant and Goswami (1995), the funds locked up in these firms will be almost impossible to recover. In many cases the financial institutions may be required to make even larger loans to these firms and hope that the firm this time around is able to become financially healthy.

Comparing the cement and steel industries there exists a distinct difference as regards the sustainability of the reforms. In cement there is absolutely no possibility of the reform process being rolled back. This is primarily because the financially distressed firms are generally speaking, small firms and moreover the exposure of financial institutions to this industry is not very large. All the major capacity expansion in this industry, when it occurs, is likely to be through the setting up of large plants (by large firms) whose financing should not be a problem at all. The future of the small cement firms looks bleak with respect to new investments.

In the case of steel however sustaining the reforms is going to be difficult. This is because financial distress is more evenly distributed across size classes in this industry and the financial institutions' exposure to this industry is also quite large. If the real prices of steel remain low for the next two years or so the situation is likely to become critical. The only way to ensure survival would be to raise import tariffs and ask the financial institutions to lend more money to this sector to fund investments that would help increase productivity.

If the financial institutions refuse to lend, the government would face a dilemma. In such a situation financial sector reforms would also have to be rolled back in the sense that the freedom granted to financial institutions in their lending operations in the post-reform period may have to be revoked. If the real prices of steel rise, then the situation may improve though the problem of funding of investments in this sector would remain.

## REFERENCES

Anant, T.C.A. and O. Goswami, 'Getting Everything Wrong: India's Policies Regarding Sick Firms', in D. Mookherjee (ed.), *Indian Industry: Policies and Performance*, New Delhi: Oxford University Press, 1995, pp. 236-88.

Gokarn, S. and R.R. Vaidya, 'Deregulation and Industrial Performance: The Indian Cement Industry' , *The Economic and Political Weekly*, 20-7 February 1993, pp. M33-M41.

———, *The Cement Industry in India: Competitive Structure, Profitability, and Prospects*, ICRA Sector Focus Series No. 3, 1994.

*Markets and Market Shares*, Economic Intelligence Service, Bombay: Centre for Monitoring the Indian Economy, various issues.

Sen, K. and R.R. Vaidya, *The Process of Financial Liberalisation in India*, New Delhi: Oxford University Press, 1997.

Sengupta, R., *The Indian Steel Industry: Investment Issues and Prospects*, ICRA Sector Focus Series No. 4, 1995.

Sidhu, S.S., *The Steel Industry in India: Problems and Perspective*, New Delhi: Vikas Publishing House Pvt. Ltd., 1987.

# Consumer Non-Durable Goods Industries: Growing Market and Increasing Competition

TIRTHANKAR ROY

In a partial way from the mid-1980s, and in a more general way from 1991-2, economic reforms were initiated in India with the explicit aim of improving efficiency. It was a partial and half-hearted effort. It introduced major changes in the spheres of foreign trade, regulation of competition, and the financial sector. But in key areas, inertia continued. Also, the reformist drive weakened quickly under three critical reverses that coincided around 1995-8: the brief return of left-of-centre parties in power, coalition politics thereafter, and severe recession for two years in exports and industry in 1996-7. Debates about how beneficial the reforms proved for poverty and welfare still continue.

This essay studies the impact of the economic reforms on the Consumer Non-Durable goods industries (CND hereafter). It will be shown here that given the very nature of the reforms, it is logical to expect that the reforms would affect these industries deeply. Further, it will be shown that the net outcome of the reforms for CND in the organized sector was, in fact, positive and significant. However, the major areas of inertia or sluggishness also led to serious problems. And the unorganized sector experience is a mixed one, a mingling of stagnation and restructuring.

The paper has five sections. Section I presents an overview of the reform process and suggests which elements in the overall picture are relevant for understanding the experience of CND. Section II describes the Annual Survey of Industries data, or the data for the registered factories. This is called the 'organized sector'. Section III describes the National Sample Survey data for unregistered manufacturing, or the 'unorganized sector'. This data is available for a small part of the period in question, but some broad inferences can be suggested based on that limited data. Section IV explores how one major source of growth, export opportunity, has transformed two large CND sectors namely clothing and leather manufacture. Section V concludes.

## I. REFORMS AND CND

### *Areas of Change and Continuity*

The logical starting point for this essay is the question: Why *a priori* one should expect the economic reform to significantly affect CND. The key elements of economic reforms relevant for CND were the following:

- First, relative factor-prices changed mainly via trade policy reforms (making the exchange rate market-determined, reduction in tariffs, and partial withdrawal of quotas). Given that India has an abundance of labour, one would expect the world prices of Indian labour-intensive manufactures to fall, and become competitive, as a result of the devaluation that followed. Given that India has a relative scarcity of capital and was technologically backward, imported capital-intensive manufacture should have become competitive due to the tariff reduction. Devaluation partly taxed that price reduction, but only partly. Withdrawal of quotas substantially reduced the transaction costs of imports of items formerly subject to licensing.
- Second, domestic deregulation in respect of licensing and direct fiscal support to specific sectors made it easier for firms to utilize economies of scale, and reduced start-up costs.

The first set of steps should generally help the CND sector, because CND by and large produces goods that tend to be very labour-intensive. The net expected effect should be an increase in exports, rise in employment, and rise in output and value-added. The second set of steps should have a mixed effect on CND. In most sectors, regulation was either excessive or perverse when policy reforms began. Deregulation, therefore, was a generally favourable step for growth. But in some cases, competition within industries between large- and small-scale, and between more and less capital-intensive firms, intensified as a result of deregulation. Where economies of scale can be significantly large, the large firm can do well. In textile production, the effect of deregulation (following the 1985 Textile Policy and subsequent changes) led to a very rapid decline in the handloom industry in competition with more capital-intensive firms. The net expected effect should be some decline in employment, but possibly an increase in productivity, as more efficient firms come in. In the net, one can expect the reforms to produce positive trends in employment, value-added, and labour productivity.

As against these changes, there were also major areas where reforms were needed, but did not happen rapidly enough. These areas of continuity mattered for the CND. The most important areas of slow change are:

- Quota restrictions continued on many types of consumer goods imports. These restrictions are expected to ease from the middle of 2001. Already, there are signs that when these ease, competition between India and other major developing manufacturing countries such as China will intensify in a wide range of goods.
- A large number of CND industries are covered by the policy of reservation for the designated small firm. This measure has created both direct regulations and indirect perverse incentives that discourage small firms from expanding in a legal and above-board way, and prevent larger firms from entering the reserved areas. In short, it has created barriers to utilization of economies of scale in production. Limited exemptions have been granted only in the case of export-oriented units, and marketing firms. These down-the-scale biases have led to poor technological effort, poor quality-control, and poor brand-creation in CNDs covered by reservation. There is reason to believe that the domestic market would have grown much faster for CND without these biases. In the last year or so, imported Chinese CNDs have made significant inroads into Indian markets, at least partly due to the cost and quality differences between Indian and Chinese goods that derive from scale factors. Recently, the new Textile Policy (2000) has de-reserved garments under pressure of sustaining competitiveness in the most important manufactured exports from India. This move is a welcome one, but it underscores the fact that de-reservation on a general scale is still unpalatable, whereas it is politically

feasible only in a sector-specific policy under special circumstances.

- In developing economies generally, export success tends to be correlated with a regime that is liberal and friendly towards foreign investors. This correlation arises from the fact that many exportables that are made cheaply in small local firms require larger and global marketing firms to reach foreign buyers. There are strong marketing economies of scale in goods that have small production economies. These marketing economies arise from a number of factors such as access to information about final markets, access to marketing infrastructure, investment on brand creation and quality control, etc. In India, a hesitant and slow-moving foreign investment policy has created barriers to such developments in global marketing. The export potential of CND could be larger with a more open FDI policy.
- Finally, infrastructure and public goods are generally scarce or of poor quality in CND. CND tends to be manufactured in small, and therefore individually vulnerable, firms. Traditionally, small firms deal with poor infrastructure and public goods by huddling together to form 'networks' that share some of these costs or provide collective solutions to common problems. But not all problems can be solved that way. Many CND firms heavily pollute air or water, but have little means to treat air or water. In CNDs such as textile printing or leather, pollution and environmental hazards have grown to become major issues in the late-1990s. No collective solution is yet in sight to deal with these. Many CND clusters are located in old craft towns that tend to be badly served by rail, road, power, water, and ports.

*The Net Outcome*

What has been the net effect of these changes and continuities on CND in recent times? It is now well-known and indisputable that:

- At least 3-4 years in the early part of the 1990s in organized sector industry in India saw historically unprecedented growth rate, by almost every index except perhaps real wages.
- Export growth also touched historically unprecedented rates from the middle of the 1980s.

- The growth phase ended in a severe slowdown the exact reasons for which are still debated. Specifically, it is not clear whether the slowdown was an effect of external forces (such as the Asian crisis, or political uncertainty in India), or due to the reform process slowing down, or due to some internal logic of the reform process itself.

The CND growth profile, as far as the registered factories are concerned, follows this general dynamics closely. There was acceleration in employment, income, output, investment, and productivity. The periods of acceleration in all these indices did not, however, exactly coincide.

As against this, there are two significant negative trends that need mention. First, all these findings are found to hold only for the organized sector. In CND, the unorganized sector is larger than the organized sector. This is an expected feature, given the labour-intensive nature of much of CND manufacture, and consequently, small economies of scale in the latter. For the unorganized sector, the picture that we get from NSS surveys, for what they are worth, is generally depressing or at best not indicative of any major change. In this respect too, CND experience and overall experience roughly tally. Second, increasingly in the course of export growth, long-term problems of foreign trade transactions with India came to the fore. These problems range from poor infrastructure to severe information problems that lead to poor quality, delivery and standardization, and to frequent frauds. Some of these transactions costs for foreign investors and buyers could have been reduced with higher firm-size, and more visible and larger partners. But firm size evolution, the reservation policy has generally discouraged.

### *CND and Other Sectors*

In respect of the analysis of growth, this study on CND shares lessons and methods with the other sectoral studies. However, in terms of specific issues, there are several differences, imposed by the subject matter. In other words, the channels by which the reforms reshaped industry differed between sectors. Some of the main differences are:

- Issues of financing growth or restructuring, especially references to the banks or large financial institutions or financial markets, do not figure in this study in a major way, because CND has rather weak links with the organized financial system, further weakened after the informal and partial withdrawal of priority sector lending schemes.

- Foreign direct investment again does not figure in this essay, while it is a major issue elsewhere.
- The sector being dominated by small firms, concentration, firm strategy, and generally speaking firm-level decision issues are relatively minor in this essay.
- On the other hand, trends in the unorganized sector are a major issue in this essay for reasons already discussed.

## II. TRENDS IN THE ORGANIZED SECTOR OF CND

### *Dataset*

The entire analysis of Section II will be based on the ASI, Factory Sector, data.[1] That is, the organized sector is defined as factories registered as such. The 'unorganized sector', described in the next section, consists of manufacturing units excluding factories registered under the Factory Act, and thus, excluding units subject to some form of regulation on employment.

### *Separation of Cotton Textiles*

In presentation of this data below, we make a distinction between cotton textiles and the rest of CND. Cotton textiles consist of manufacture and processing of cloth. There are two obvious grounds for treating this sector separately:

- It is the largest subsector within CND and worth looking at separately.
- A large part of this subsector produces an output—cloth—that can be used as both consumption good and as input.

Cloth is hard to define as a final consumption good, or an intermediate. In reality, this sector performs both these roles, it sells cloth to consumers directly as garments or as pieces to be stitched into garments, and it also sells inputs to the clothing sector.

There is a third and less obvious reason for treating cotton textiles separately. While CND generally has not been subject to significant extent of direct regulation, cotton textiles was an exception, and was subject to a wide range of regulatory policies that were aimed at restraining competition between handloom weaving and the cotton mills. This regime was partially rolled back in 1985, but by then it had left a deep impact on the industry. In particular, while the handlooms continued to decline due to inevitable technological obsolescence, the majority of the mills too had become irretrievably bankrupt due to lack of expansion, diversification, and modernization. It is due to the accumulated effects of this regime, and the decline of the handlooms, that cotton textiles come out as a significant exception to the general picture of growth.

*Periodization*

Of the major changes central to the reform package, several were effective before 1991-2, the date when reforms became an official agenda. Liberalization of the exchange rate began slowly from the early 1980s, with immediate effect on exports of CND. Industry-specific steps towards deregulation began in the mid-1980s. One of the most important of these for the present essay was the Textile Policy of 1985, which did away with capacity restrictions on the textile mills. On the other hand, tariff reduction and quota withdrawal were measures that came much later.

When the reform period really begins is, therefore, a question that needs an answer before we can compare performance between periods. Did the reforms begin in 1985 or in 1991? One way to answer this question would be to see whether statistically 1985 or 1991 serves as the more robust cut-off point.

Growth rates in employment, real income, real output, real investment and labour productivity for a few years after 1991-2 tend to be so consistently large that year-dummies with values of 1 for 1991 or after and 0 otherwise, give much better statistics than values of 1 for 1985 and after and 0 otherwise. However, as Figure 1 shows, 1985 was evidently a turnaround point in employment. That it was so can be tested statistically with an F-test for structural break at 1985. The test confirms strongly that 1985 was indeed a structural break for employment. As for the other variables, 1985 does not appear to be a decisive break in a statistical sense, as can be expected from looking at Figure 2.

Ultimately, the precise cut-off date does not seem to matter very much, because whether we use 1985 or 1991/2, the same result comes through, that CND growth rates accelerated in the second period compared with the first. Table 1 shows this clearly. The table also shows that the acceleration probably began sometime in the second half of the 1980s, since adding these years improves growth rates generally, with the exception of employment.

In the rest of this section, 1985 is taken as the standard cut-off point.

*Employment, Wage, Productivity*

Tables 2 and 4 show that employment growth rate accelerated in CND from the late 1980s, in significant contrast with industry as a whole. In cotton textiles there was a deceleration, so if textiles is excluded, the increase in growth rate in CND appears more impressive. With this shift, the share of CND in all industries has increased too, reversing a decline in the pre-reform period. In other words, the reform process helped CND especially. This can be understood in terms of one of the objectives of trade and domestic market deregulation, to allow comparative advantage to work. CND is intensive in a surplus resource—manual, informally trained labour—and thus has generally done well both at home and abroad. Excluding cotton textiles, CND share in all industries has increased, which confirms (a) the positive impact of the reforms on labour-intensive industry, (b) the positive impact of the reforms on consumption expenditure.

TABLE 1. MAJOR RATIOS FOR THE ORGANIZED SECTOR (*% per annum*)

| | *Trend growth rates* | | | | *Gross fixed capital formation as a ratio of gross output* |
|---|---|---|---|---|---|
| | *Employment* | *Real value-added* | *Real output* | *Real productivity* | |
| 1975-6 to 1988-9 | 0.0 | 5.3 | 5.0 | 5.3 | 4.4 |
| 1990-1 to 1997-8 | 3.0 | 8.3 | 7.1 | 5.3 | 6.2 |
| 1975-6 to 1984-5 | 1.3 | 4.9 | 4.6 | 3.6 | 4.4 |
| 1985-6 to 1997-8 | 2.6 | 8.4 | 7.9 | 5.6 | 5.6 |

NOTE: The year 1989-90 saw a large drop in value indices and has been excluded from the first set of figures. For the last column, data start from 1980-1.

The acceleration in employment growth rate seems to be a general feature of organized manufacturing if 1991 is used as the dividing line rather than 1985. There is now a large literature dealing with the markedly slow employment growth of the 1980s, part of it compares the 1980s with the acceleration in the 1990s. The explanation hinges on how trends in real wages are interpreted. Real wages grew quite rapidly in the 1980s and sluggishly in the 1990s. In one view, employers in the 1980s adopted labour-displacing technologies, which led to a slow employment growth but rapid wage growth. In another view, the rapid wage growth was a reflection of labour politics, and employers were looking for flexibility via subcontracting or displacement.[2] These two theories—employers substituted labour or yielded to labour—cannot be easily tested because of identification problems. The available data is too aggregated to suggest motives, as such it can be interpreted with equal plausibility in either way. In any case, the 1990s appear as a period when, by contrast, employment growth rate increased and wage growth rate fell due to either political weakness of labour or the increasing importance of industries intensive in semi-skilled labour. CND was the major component among such industries.

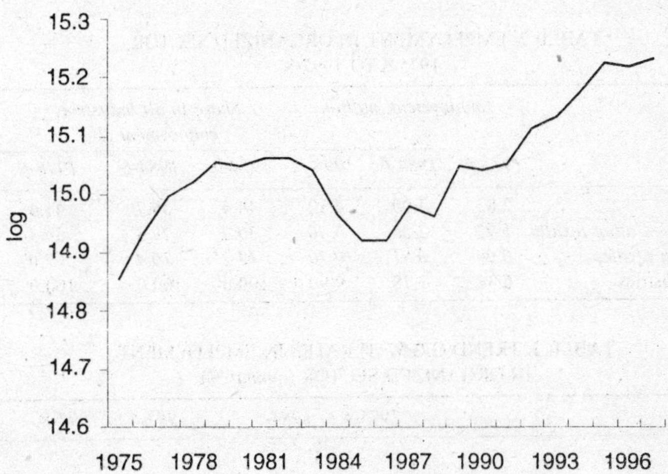


FIGURE 1. EMPLOYMENT IN CND

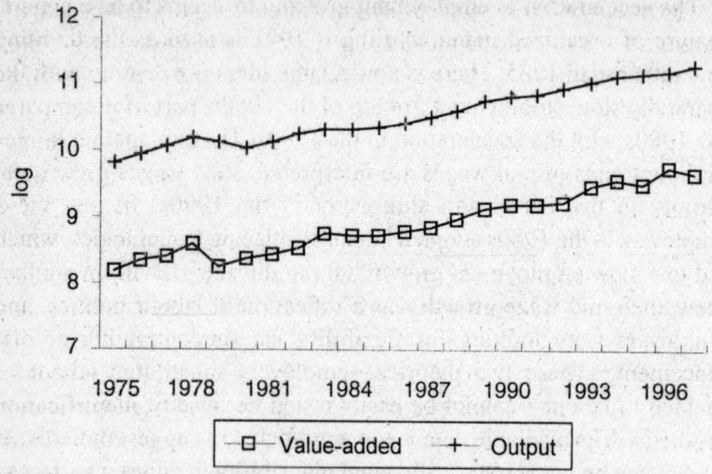


FIGURE 2. INCOME AND OUTPUT IN CND (1981-2 PRICES)

The reform process is also expected to have intensified competition and enabled import of technology. These measures are expected to raise labour productivity. On the other hand, the implicit non-interference of the government in the operations of the labour market suggests that wages became progressively market-driven. Given the

TABLE 2. EMPLOYMENT IN ORGANIZED SECTOR, 1975-6 TO 1997-8

| | *Employment, millions* | | | *Share in all industries employment (%)* | | |
|---|---|---|---|---|---|---|
| | *1975-6* | *1984-6* | *1996-8* | *1974-6* | *1984-6* | *1996-8* |
| CND | 2.83 | 3.09 | 4.10 | 44.4 | 39.7 | 41.4 |
| *CND – Cotton textiles* | *1.93* | *2.28* | *3.40* | *30.2* | *29.3* | *34.3* |
| *Cotton textiles* | *0.90* | *0.81* | *0.70* | *14.2* | *10.4* | *7.0* |
| All industries | 6.38 | 7.78 | 9.90 | 100.0 | 100.0 | 100.0 |

TABLE 3. TREND GROWTH RATES IN EMPLOYMENT IN ORGANIZED SECTOR (*annual %*)

| | *1975-6 to 1984-5* | *1985-6 to 1997-8* |
|---|---|---|
| CND | 1.3 | 2.6 |
| *CND – Cotton textiles* | *1.9* | *3.8* |
| *Cotton textiles* | *- 0.1* | *- 0.7* |
| All industries | 2.6 | 2.3 |

existence of surplus labour, the prediction is that the growth rate of real wages would slow down in the reform period. These predictions are tested in Tables 4-6. These three tables lead to the following general conclusions:

- *In CND, growth in productivity accelerated.* This is a general feature of industry as a whole (Tables 4 and 6).
- *There was a reversal in CND's position relative to that of all industries.* A rapid declining trend in CND's relative productivity was arrested from the mid-1980s.

TABLE 4. LABOUR PRODUCTIVITY (AT 1981-2 PRICES) IN THE ORGANIZED SECTOR

| | *Productivity in real terms (value-added, Rs/year/person at 1981-2 prices)* | | | *Proportion of all industries productivity (%)* | | |
|---|---|---|---|---|---|---|
| | *1974-6* | *1984-6* | *1996-8* | *1974-6* | *1984-6* | *1996-8* |
| CND | 13111 | 19883 | 39524 | 81.5 | 68.7 | 67.6 |
| *CND – Cotton textiles* | *13217* | *20942* | *41457* | *82.1* | *72.3* | *70.9* |
| *Cotton textiles* | *12883* | *16892* | *30099* | *80.1* | *58.3* | *51.5* |
| All industries | 16093 | 28949 | 58456 | 100.0 | 100.0 | 100.0 |

TABLE 5. REAL WAGES IN THE ORGANIZED SECTOR

| | *Real wages (Rs/year/person)* | | | *Proportion of all industries wages (%)* | | | *Wage-share in value-added* | | |
|---|---|---|---|---|---|---|---|---|---|
| | *1975-6* | *1984-6* | *1996-8* | *1975-6* | *1984-6* | *1996-8* | *1975-6* | *1984-6* | *1996-8* |
| CND | 6196 | 8030 | 8714 | 78.8 | 75.8 | 66.6 | 52.2 | 45.4 | 27.6 |
| *CND – Cotton textiles* | *4941* | *7222* | *8363* | *62.8* | *68.1* | *63.9* | *41.2* | *38.8* | 25.4 |
| *Cotton textiles* | *8876* | *10236* | *10426* | *112.8* | *96.6* | *79.6* | *76.4* | *68.3* | 42.5 |
| All industries | 7865 | 10519 | 13091 | 100.0 | 100.0 | 100.0 | 54.2 | 41.3 | 27.6 |

TABLE 6. TREND GROWTH RATES IN REAL WAGES, PRODUCTIVITY AND CAPITAL PER WORKER

| | *Wage* | | *Productivity* | | *Capital per worker* | |
|---|---|---|---|---|---|---|
| | *I* | *II* | *I* | *II* | *I** | *II* |
| CND | 2.4 | 0.8 | 3.6 | 5.6 | 4.8 | 9.6 |
| *CND – Cotton textiles* | *3.9* | *1.3* | *4.5* | *5.6* | *4.2* | *9.3* |
| *Cotton textiles* | *0.7* | *0.2* | *1.2* | *4.3* | *6.3* | *11.5* |
| All industries | 2.6 | 1.6 | 5.0 | 6.6 | 7.7 | 10.0 |

NOTES: * Point-to-point growth rates between 1979-80 and 1984-5. Period I refers to 1975-6 to 1984-5, and Period II refers to 1984-5 to 1997-8.

- *In all sectors, rise in real wages slowed down.* As we have seen, the explanation for the simultaneous occurrence of a slowdown in wage-growth and rise in employment-growth is likely to be controversial. Two relevant facts are not likely to be disputable. First, in the 1990s, the incidence of industrial disputes has come down to a remarkable extent, which suggests, however indirectly, changes in the politics of labour that might encourage more employment. And second, macroeconomic policy has favoured exportables generally, which do tend to be intensive in unskilled labour, which might again explain the joint occurrence of deceleration in wages and acceleration in employment. Implicit in the second formulation, is the idea that there is persistent oversupply in the market for unskilled labour. But whether politics or economics that should be the main explanatory variable remains to be further researched.
- *Wage-share in value-added fell* universally much faster in Period 2. Growth of non-wage incomes in real terms accelerated.
- *Industry capitalized much faster* (Table 6), which must partially account for the rise in productivity.

In short, the growth in employment in the second period seems to have been driven by changes in input, especially labour markets, and by easier access to technology.

The inference that slowdown in wage-growth due to political or economic reasons encouraged acceleration in employment growth needs to be qualified. There is ground to believe in such a logic only over time. There is a negative, but not significant, correlation between wage-growth (first difference of log) and employment-growth in the reform period. But across industries, the correlation is very weak, though still of the correct sign. It would seem that over time wage-growth has slowed down, but across industries, labour demand has had a positive influence on wage-growth.

*Investment*

Rates of investment in organized sector CND shows a very striking pattern. As Figure 3 shows, investment-output ratio rose steadily throughout the post-1985 period, to sharply accelerate in 1991-5. Thereafter, an equally sharp decline took place. It is easy to explain an investment boom by at least five factors:

- The exceptional trends in profitability from the mid-1980s (Fig. 4).
- The absence of sharp recessions in the first half of the 1990s.
- Case studies on specific sectors and studies on investment trends have suggested that along with expected returns, availability of low-cost funds was a major reason for the investment boom in industry. The stock market boom in 1992-5, the inflow of foreign portfolio and direct investment, and the first wave of bank deregulation, further eased financial flows.
- Deregulation of capacity expansion played a role.
- Deregulation of technology import very clearly played a role in such CNDs as textile products, food products, etc.

A good illustration of all five factors is the cotton textile industry, where there was an improvement in investment rates in the 1990s. I have explored elsewhere how deregulation in textiles after the 1985 policy speeded up modernization in a few well-managed companies among the largely bankrupt integrated cotton mills, and the 1991 reforms and the Export Promotion Capital Goods scheme further aided this process. Finance for this modernization came from yarn exports in the 1980s, and later from equity (1992-4, both big and small companies) and debentures (until 1992, only for the bigger companies). The dozen or so modernizing mills diversified away from generic goods, from cloths where powerloom competition could be serious and which offered no particular economy of scale or integration, and specialized in certain high-quality cloths where economy of scale and integration of spinning-weaving-processing was significant. These mills simultaneously exported high-quality cloth, and supplied them at home through non-traditional retailing systems.[3]

Supply of funds can also explain the severity of the fall in investment rate in the second half of the 1990s. Profitability declined, but not dramatically. On the other hand, real interest rates rose, and the equity market more or less dried up after 1995 for most new projects.

### *Size and Location*

A long-term tendency of declining average size of firms (workers per unit) was arrested in the post-reform period. This is in contrast with the average picture where size continued to decline in the 1990s.

However, as we can infer from the data on value-added growth, firms have increased in size in terms of income generated. In other words, there was no real decline in firm size, but workers produced more, via technological and institutional means.

In industrial composition, there has been no major diversification in the long period (Table 7). Food and textiles dominate CND sector. However, within textiles, there has been a shift towards downstream products and towards non-cotton. At the same time, cotton intermediate manufacture has shifted out of the large-scale to small-scale sphere. The major sectors that increased in share were—other textiles, food, beverages and tobacco, drugs and cosmetics, and leather. This list immediately suggests one powerful external force behind growth in markets—export opportunities. All these groups were driven to greater or lesser extent by the prospect of exports. However, another discernible demand factor behind changes in industrial composition was preference in the home market. Consumption of higher-valued items and plastics increased. This is reflected in the rising shares of smaller groups such as batteries (showing increasing consumption of electrical and electronic goods) and plasticware, both included in 'other CND' in Table 7.

While employment growth rate has increased in CND in the 1990s, there has been a tendency towards ruralization of CND as well as overall employment. This tendency can be estimated only approximately because rural-urban break-up at 3-digit level is not available. Using eight 2-digit groups that contain mainly CND goods, there seems to be an overall ruralization, that is, increase in the share of

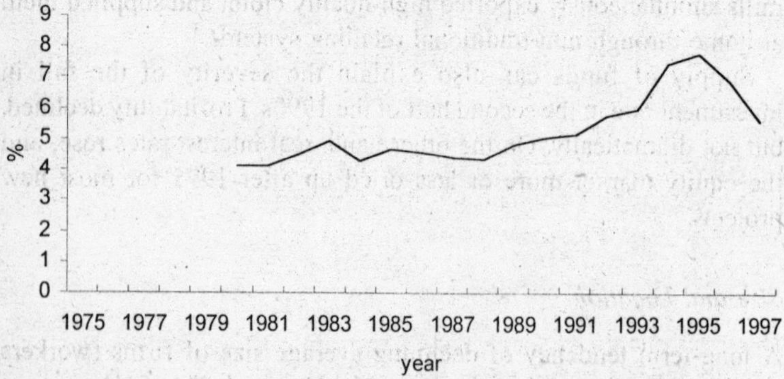


FIGURE 3. RATES OF INVESTMENT (*Investment-output ratio, %*)

TABLE 7. INDUSTRIAL COMPOSITION IN THE ORGANIZED SECTOR, 1975-6 TO 1997-8

| | *Share in all industries employment (%)* | | |
|---|---|---|---|
| | *1975-6* | *1984-6* | *1996-8* |
| CND | 44.4 | 39.7 | 41.4 |
| CND – cotton textiles | 30.2 | 29.3 | 34.3 |
| All industries | 100.0 | 100.0 | 100.0 |
| *Major groups* | | | |
| Food and beverages | 19.9 | 17.5 | 19.4 |
| Textiles | 18.0 | 15.4 | 13.2 |
| *Detailed groups* | | | |
| Cotton textiles | 14.2 | 10.4 | 7.1 |
| Other textiles and textile products | 3.8 | 5.0 | 7.1 |
| Leather | 0.1 | 0.5 | 0.9 |
| Paper | 2.3 | 2.1 | 1.7 |
| Drugs and cosmetics | 1.7 | 2.0 | 2.6 |
| Other CND | 2.6 | 2.3 | 2.5 |

rural factories in employment. But individual industry experiences differ somewhat. There has been very sharp ruralization in group 23 (cotton textiles), which reflects the decline of the cotton mills, which are located in the cities. There has also been a sharp ruralization in wood products, which may reflect a move towards sources of raw material. On the other hand, in food products (20-1, 22), there has been no sharp change. And in groups 26 (textile products) and 29 (leather), there has been urbanization. Significantly, both groups are export-oriented, and thus possibly more dependent on urban infrastruc-

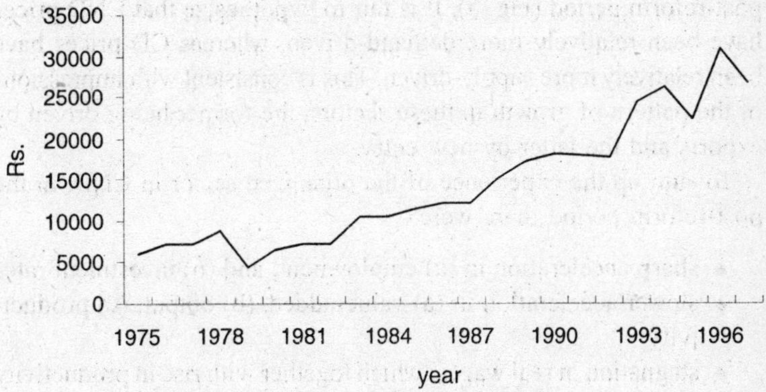


FIGURE 4. PROFIT PER WORKER (*1981-2 prices*)

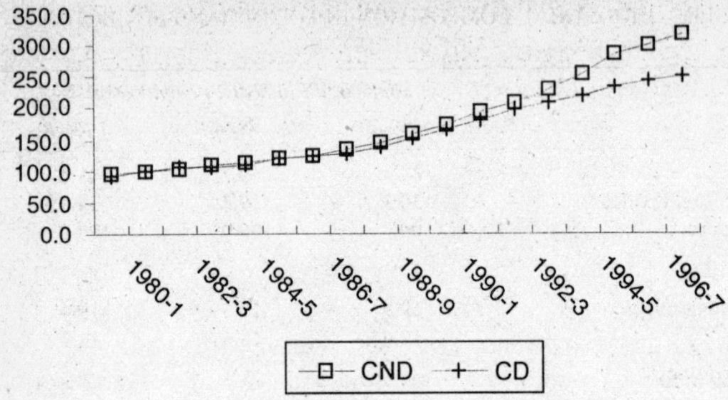


FIGURE 5. PRICES

ture. Leaving aside industry-specific factors, it is worth investigating if there is some universal reason why CND may be ruralizing. Partly, this tendency may be spurious, in that it may only reflect growth in industry in suburbs of towns away from the core areas where land prices and rents are going up. If not entirely spurious, the tendency may be driven by (a) diversification in rural consumption, (b) more dispersed origin of investible capital in industry, and (c) increase in growth of rural consumption.

*Prices*

About trends in prices, there is only point that needs to be noted. Prices of CND and consumer durables (CD) have diverged in the post-reform period (Fig. 5). It is fair to hypothesize that CND prices have been relatively more demand-driven, whereas CD prices have been relatively more supply-driven. This is consistent with impressions of the pattern of growth in these sectors, the former being driven by exports and the latter by new entry.

To sum up the experience of the organized sector in CND, in the post-reform period there were:

- sharp acceleration in (a) employment, and (b) investment rates
- slower acceleration in (a) value-added, (b) output, (c) productivity
- stagnation in real wages, which together with rise in productivity implies increase in profitability. Figure 3 confirms that impression.

In short, the 1990s were a period of growth—not unusual in extent, but unusually profitable—it was driven by sectors where entry was easy and wages were driven by demand and supply rather than by unionization and regulation.

In several respects, the experience of the unorganized sector in CND has been a stark contrast.

## III. TRENDS IN THE UNORGANIZED SECTOR OF CND

The 45th and 51st rounds of the National Sample Survey Organization (NSSO) conducted sample surveys of the 'unorganized' manufacturing sector.[4] These were done in 1989-90 and 1994-5. Unorganized manufacturing covers units that are not registered under the Factory Act. The Economic Census, 1980 and 1990, also collected detailed data for small-scale manufacturing establishments, but the Economic Census is not always comparable with the NSS. The former used the house-list of the 1981 and the 1991 population censuses, its employment figures come close to those of the censuses, but between the census and the NSS there are big differences in employment estimates. So, we decided to avoid using the Economic Census totals.

Between the population census and the NSS surveys, there occurs a systematic difference in levels of employment. The NSS figures are much higher. The difference is the widest for estimates of women industrial workers, especially wide for urban women. *A priori*, one would believe that the population census, being a census, is the more reliable of the two. However, many scholars dealing with employment think the NSS gives a more accurate picture, and that the census underestimates women workers.[5] This is possible, for the census tends to use a stricter definition of workers that sometimes leaves out part-time workers, and many women tend to be part-time workers. Still, precisely why the census operations cause this discrepancy is not absolutely clear. Nevertheless, the census admits that there is a problem, which shows up in the slogan 'Women also Work', flashing in the census 2001 web-site.

We need to note first of all that CND employment is much larger in the unorganized sector (17.5 million in 1994-5) than in the organized sector (4 million in 1996-8). On the other hand, productivity is much higher in the latter, almost as much as 10 times higher. This makes one suspect that the NSS uses rather too liberal a definition of

TABLE 8. EMPLOYMENT IN UNORGANIZED SECTOR, 1989-90 AND 1994-5

| | *Employment, millions* | | *Share in all industries employment (%)* | |
|---|---|---|---|---|
| | *1989-90* | *1994-5* | *1989-90* | *1994-5* |
| CND | 19.34 | 17.47 | 54.5 | 52.6 |
| *CND – Cotton textiles* | *16.69* | *15.82* | *47.0* | *47.7* |
| *Cotton textiles* | *2.65* | *1.64* | *7.5* | *5.0* |
| All industries | 35.47 | 33.20 | 100.0 | 100.0 |

TABLE 9. REAL VALUE-ADDED AND OUTPUT IN THE UNORGANIZED SECTOR, 1989-90 AND 1994-5

| | *Value-added in 1981-2 prices (Rs billion)* | | *Output in 1981-2 prices (Rs billion)* | |
|---|---|---|---|---|
| | *1989-90* | *1994-5* | *1989-90* | *1994-5* |
| CND | 59.49 | 58.96 | 186.85 | 155.58 |
| *CND – Cotton textiles* | 54.52 | 54.09 | 163.96 | 140.61 |
| *Cotton textiles* | 6.98 | 4.87 | 22.89 | 14.97 |
| All industries | 117.79 | 126.17 | 337.98 | 312.30 |

TABLE 10. LABOUR PRODUCTIVITY (AT 1981-2 PRICES) IN THE UNORGANIZED SECTOR

| | *1989-90* | *1994-5* |
|---|---|---|
| CND | 3077 | 3376 |
| *CND – Cotton textiles* | 3147 | 3418 |
| *Cotton textiles* | 2636 | 2964 |
| All industries | 3320 | 3800 |

TABLE 11. PERCENTAGE CHANGE BETWEEN 1989-90 AND 1994-5

| | | *Employment* | *Productivity* |
|---|---|---|---|
| CND | Unorganized | – 9.7 | + 9.7 |
| | Organized | + 13.3 | + 34.7 |
| All Industries | Unorganized | – 6.4 | + 14.4 |
| | Organized | + 11.8 | + 63.6 |

'worker' in its surveys. At any rate, the average productivity and wages in the unorganized sector are no indication of notional full-time earnings and productivity. For, a very large percentage of these workers are part-timers (22 in 1994-5) and unpaid (34).

Table 8 summarizes the employment situation in the 1990s. In CND and in the unorganized sector as a whole, employment declined in the first half of the 1990s. This experience is in sharp contrast with that of the organized sector. Table 11 brings out the contrast clearly. On the other hand, real output and real value-added changed to a smaller extent in CND and the unorganized sector (Tables 9, 10). As a net result, the CND sector as well as the unorganized sector as a whole experienced a significant rise in productivity that is comparable with the rise in the organized sector.

What can we conclude from this mixed picture? First of all, there must exist large segments within the unorganized sector CND that have lost their markets. These segments have contributed to the fall in income and output, as well as employment. But that cannot be the only explanation for the decline in employment, because productivity has increased on average. Why would anyone be encouraged to increase efficiency or invest if the market is doomed? Clearly, in large segments, not only has there been an *expansion*, but also signs of improvement in efficiency. Only such workers have become unemployed who had low levels of productivity. These signs are not consistent with a picture of general decline. Indeed a picture of general decline in the unorganized sector CND will come as a great surprise, for it can be expected that this sector consists of some fast-growing exportables.

A closer look at industrial composition (Table 12) helps us narrow down on the sectors that experienced large fall in employment. The three main conclusions on compositional change are:

- Decline of textiles—driven predominantly by obsolescence in handlooms.
- Rise of textile products and leather—driven by exports.
- Rise in paper, drugs and cosmetics, driven by domestic demand.

On some of these respects there is broad similarity with organized sector.

The leading sector in fall in employment was handloom weaving, which dominates the general textiles category. Interestingly enough, the handloom sector did experience rise in productivity.

Data on structural change show that there was sharp fall in:

- the extent of household industry (the share of Own Account Manufacturing Enterprises fell from 87 to 70), and
- rural industry (share in employment fell from 77 to 71).

TABLE 12. INDUSTRIAL COMPOSITION: UNORGANIZED SECTOR

(*percentage of employment*)

| | *1989-90* | *1994-5* |
|---|---|---|
| CND | 54.5 | 52.6 |
| CND – Cotton textiles | 47.0 | 47.7 |
| All industries | 100.0 | 100.0 |
| *Major groups* | | |
| Food and beverages | 26.2 | 25.2 |
| Textiles | 18.4 | 16.9 |
| *Detailed groups* | | |
| Cotton Textiles | 7.5 | 5.0 |
| Other textiles and textile products | 11.0 | 12.0 |
| Leather | 1.4 | 1.5 |
| Paper | 1.4 | 2.2 |
| Drugs and cosmetics | 0.6 | 0.7 |
| Other CND | 6.4 | 6.0 |

These two features are consistent with one another, since rural areas have a concentration of household industry, and are together consistent with one of the central findings from the population censuses, namely, decline of household industry. These findings are also consistent with expectations about how labour markets should behave with increasing competition and market-transactions. One would normally expect labour to commercialize, that is wage-labour to increase, when the products that they make commercialize or face intense competition. This connection works via several mechanisms. One mechanism is, more market exchange leads to keener competition for resources like capital and information, households tend to be worse off in this competition than do capitalists. Another mechanism is, markets and competition encourage producers to employ greater division of labour, and utilize economies of scale, thus forcing the units to hire in labour. Economies of agglomeration also become important advantages. These latter economies encourage migration, migrations typically involve males, and thus break up the unity of the household as a production unit. The population censuses, in fact, do confirm the working of this logic for at least as long as the period 1961-91, when household industry steadily declined. NSS too shows a similar trend.

These findings, however, are *not* consistent with another result that there was a slight fall in average size per unit, from about 3.2 to

2.5. Normally, a fall in the importance of the household industry should lead to a rise in the average size of unit in terms of employment.

## IV. THE ROLE OF EXPORT OPPORTUNITY ON CND: TWO STUDIES

Clothing and leather manufacture are the two striking examples of export-driven growth, both form major sectors within CND. Textile and leather exports grew extremely rapidly from the 1980s. Their combined share in total exports increased from 14 per cent in 1970 to 30 per cent in 1998 (Table 13). Correspondingly, their combined share in factory employment increased from 3 per cent in 1975 to 11 per cent in 1997. While this growth testifies to the success of the reforms in encouraging export of CND, it has also exposed a number of serious structural weaknesses of these industries. In the 1997-8 recession, these weaknesses were highlighted further. It is worth taking a detailed look at these two sectors to uncover some typical strengths and weaknesses of Indian CND in the world market.

### *Clothing*

India has a large, diversified and versatile base in fibres and cloths. Labour is among the cheapest here. The combination of these two advantages few garment exporters in the developing world share.

TABLE 13. TEXTILE AND LEATHER EXPORTS, 1970-98

| | *1970* | *1980* | *1985* | *1990* | *1995* | *1998* |
|---|---|---|---|---|---|---|
| Total export (Rs billion) | 15.4 | 67.1 | 109.0 | 325.5 | 1063.5 | 1416.0 |
| Export share in GDP (%) | 4 | 5 | 4 | 6 | 10 | 9* |
| *Textiles and Clothing* | | | | | | |
| Export of textiles (Rs billion) | 1.5 | 9.3 | 21.3 | 68.3 | 241.5 | 359.0 |
| Textile export in real terms (Rs billion)* | 1.5 | 3.7 | 6.1 | 13.6 | 29.0 | 37.9 |
| Textiles' share in export (%) | 9 | 14 | 20 | 21 | 23 | 25 |
| Clothing's share in textile export | 20 | 59 | 59 | 59 | 51 | 52 |
| *Leather* | | | | | | |
| Export of leather and leather goods (Rs billion) | 0.8 | 3.9 | 7.7 | 26.0 | 57.9 | 68.2 |
| Leather's share in export (%) | 5 | 6 | 7 | 8 | 5 | 5 |
| Textiles and leather exports ($ billion) | 0.3 | 1.7 | 2.1 | 5.2 | 8.9 | 10.1 |

NOTE: * Deflated by Wholesale Price Indices.
SOURCE: India, Ministry of Finance, *Economic Survey*, New Delhi, various years.

And yet, India's performance as a garment exporter was quantitatively negligible until 1980, in the context of unfriendly trade regime. Thereafter, encouraged by steady devaluation of the rupee, garment exports began to increase very rapidly. India's share in the world clothing trade also increased. In the next twenty years, the clothing industry experienced not only a dramatic growth, but also structural changes that could not have been imagined before. Some of the salient changes are:

- Improvement in manufacturing capability, measured in terms of the range of fibres and range of garments now produced compared to 20 years ago.
- Massive growth of tailoring factories supplying not consumers directly, but to clothing exporters, as well as wholesalers and retailers in the domestic market.
- The retailing and wholesaling of garments have expanded at the cost of wholesaling of fabrics.
- Simultaneously, domestic consumers have shifted in a marked way from customized garments to readymade garments.
- Clothing tastes have globalized significantly. There is far greater sensitivity and contact with international fashion trends.

Altogether, export market, via improved capability in both production and marketing, has also revolutionized domestic clothing consumption, production and marketing.

Yet, neither the export story nor the counterpart home market story, is one of unmixed success. Despite steady double-digit growth rates in exports maintained for nearly 15 years, India's performance in garment trade has been rather poor, especially by the standard of East and South-East Asia. Under the Agreement on Textiles and Clothing, WTO, India must allow import of garments soon, and it is believably held that Chinese imports will succeed in India to a large extent. What is lacking?

Two critical deficiencies of Indian clothing, by comparison are most commonly cited. These are, poor quality, and narrow range. As a result of these deficiencies, India can compete only in markets that value cost more than quality. Such markets include (a) other developing countries within easy reach from India, (b) items consumed in developed countries where cost matters relatively more. These markets do not include high-value-added garments. The low-cost image rules out India's access to the larger part of the developed country markets.

Why do these deficiencies arise? The various reasons cited can be listed as follows:

1. *Economies of scale are not fully utilized.* Indian clothing factories tend to be smaller in size than in the rest of Asia. This smallness is partly a policy problem, a result of the reservation of clothing for the small-scale sector, which has just been revoked. Partly, it is a structural problem, arising from the high cost of capital. Elsewhere in Asia, a more open FDI policy mitigated capital scarcity in exportable industries to some extent, while also enabling market access. In India, FDI policy has been considerably more rigid.
2. As a result of poor globalization of enterprises, *retailing abroad has remained undeveloped.*
3. *Garment-textile interface has been limited* in India. The textile industry includes powerlooms and large integrated mills. The latter, who have better technology on average, traditionally retailed cloth pieces rather than supplying the clothing industry. This old habit has led to the mills being very badly responsive to the garment industry's needs. On the other hand, the powerlooms are on average not equipped to supply world-class quality textiles.
4. As a result of the above factor, the *quality of cloth in India tends to be poor*, if cheap.
5. While *labour is cheap in India, it is not of very high quality.* Both managerial and technical practices tend to be deficient and wasteful in clothing manufacture. For example, tailors trained in traditional lines of work are placed in charge of cutting. The former are not usually aware of standardization requirements, which requires a different skill and mentality altogether.
6. *Infrastructure* such as ports, railways and roads are of much worse quality in India than in, say, China. This problem makes small firms bear a great deal of direct and indirect costs while exporting.

The threat of better-quality imports realized partly in 1997-8 after the recession in South-East Asia. In a more legal and open way, it is going to occur from early 2001. This impending threat has made the industry and the government redesign strategy. Under this pressure, the garment industry has recently been de-reserved, promised technological upgradation fund, and free trade zones. There is also great need for micro-level interventions such as worker-retraining, to initiate which competition will probably is the most effective instrument.

*Leather Manufacture*

World trade in leather and leather products has increased very rapidly in the last 20 years. Simultaneously, there has occurred a change in the direction of trade, with the West turning into large net importers, and the developing world large net exporters. The leather industry in India has always been a large one thanks to the abundance of its main raw materials, hides of cattle and buffalo, and skins from sheep and goats. From the 1980s, the industry benefited significantly from exchange rate deregulation and grew very rapidly. Although export-oriented, the export of manufactures itself created new capability especially in footwear manufacture that spilled over into the domestic market. In a rather similar way as clothing, domestic consumption, production and marketing of footwear changed in these 20 years. There was shift from customized footwear to buying from wholesalers and large retailers, the trade network itself expanded and diversified, effectively ending the semi-monopoly of Bata, the dominant firm in mass-produced footwear for a long time.

While Indian exports did grow, China's expanded much faster, so that India's share in world trade grew to a small extent and stands at about 3 per cent. The most important item of world trade in leather is footwear, which accounts for 40 per cent. India is a small exporter of footwear, with a share of 1 per cent against China's 14 per cent. India is more competitive in leather garments and finished leather itself.

While the basic cost conditions favour India, leather exports have faced three serious challenges in recent years. All three call for major policy reforms. The problems are: competition from new production centres, technological backwardness and quality deficiencies, and campaign of animal rights activists. The leather trade is quickly becoming more competitive with the expansion of capacity in South-East Asia, mainly Vietnam, and in Eastern Europe. In the Indian industry, reservation created the same kind of distortions and scale-biases as in garments manufacture.

During January-June 2000, a serious threat to Indian leather export arose from a totally unexpected source, animal rights activists led by the international NGO, People for the Ethical Treatment of Animals or PETA. They demonstrated with graphic video footage, extreme cruelty meted out to animals taken for slaughter. Their high-profile campaign targeted to major foreign stores succeeded when The Gap and its two associates stopped buying leather from India earlier in 2000, and several other international stores stated their intention to review buying from India. The panic-stricken industry first tried to evade the charge by shifting the responsibility on the meat industry, and alleging a political conspiracy aimed at India whereas Pakistan, China, or Bangladesh, where it is believed cruelty to animals is equally rampant, were left out. But under sustained attack, the industry and the government began to talk about concrete steps to deal with cruelty.

It is important to know why cruelty to animals is so rampant in India. Insensitiveness is only part of the issue. Cruelty arises also from hypocritical policies and their poor implementation. Cattle in India are one of the biggest foreign exchange earners. They are also an object of veneration. Consequently, while the government encourages the leather trade abroad, it holds on to regulations curbing freedom of trade in the raw material for the industry. In most states in India cow slaughter is illegal. At the same time, under India's National Cattle Act, it is illegal to take cattle across state borders for slaughter. The net outcome is a transport system of cattle run by the mafia and largely underground. Much of the cruelty the PETA documented arises from the need to remain clandestine while taking cattle across borders. Under the current political climate, such contradictions cannot be resolved. Therefore, the suggested measures to deal with cruelty do not touch on the issue of increasing freedom of trade by deregulation in key areas. On the other hand, what other measures can be taken is seriously compromised by the semi-legal nature of the trade.

## V. CONCLUSION

To conclude, four main points need to be restated.

1. In the organized sector, the post-reform period has seen a distinct improvement in performance and efficiency.

2. The unorganized sector presents a more complex picture. If we confine ourselves to employment, income or output, it is a picture of decline. If we look at labour productivity, unorganized sector has increased in efficiency just as impressively as the organized sector. It is too simplistic to suggest that the unorganized sector as a whole declined in the post-reform period. The handloom industry saw the largest decline in employment, and yet saw increase in productivity per worker, which is not consistent with a fall in markets. A more sober conclusion will be that segments of unorganized industry saw shrinking markets, whereas other segments saw adjustments to improve efficiency and competitiveness.
3. The unorganized sector also experienced urbanization and increasing employment of wage-labour in place of family labour.
4. Export opportunities came as a boon for specific segments inside organized sector as well as unorganized sector CND. But the experience of exporting also revealed serious weaknesses of the supplier firms, weaknesses that arose partly due to inexperience, and partly due to absence of regulation.

## ACKNOWLEDGEMENTS

This paper benefited greatly from comments and suggestions received from Shuji Uchikawa on successive drafts, and from response of other members of the team during a meeting where the findings of the papers were discussed.

## NOTES

1. Central Statistical Organisation (CSO), *Annual Survey of Industries*, New Delhi.
2. For a selection of writings interpreting the 1980s 'jobless growth', I.J. Ahluwalia, *Productivity and Growth in Indian Manufacturing*, New Delhi: Oxford University Press, 1992; Ajit Ghose, 'Economic Restructuring, Employment and Safety Nets', ILO-ARTEP, *Social Dimensions of Structural Adjustment in India*, Delhi: ILO, 1995; R. Nagraj, 'Employment and Growth in Manufacturing Industries: Trends, Hypotheses and Evidence', *Economic and Political Weekly*, 29(4), 27 January 1994; S.R. Bhalotra, 'The Puzzle of Jobless Growth in Indian Manufacturing', *Oxford Bulletin of Economics and Statistics*, 60(1), 1998; K.V. Ramaswamy, 'The Search for Flexibility in Indian Manufacturing: New Evidence on Outsourcing

Activities', *Economic and Political Weekly*, 34, 6-12 February 1999. For a survey of this literature and an interpretation of the 1990s, see B.N. Goldar, 'Employment Growth in Organized Manufacturing Growth in India', *Economic and Political Weekly,* 35(14), 1 April 2000. On labour issues specifically see Dabashish Bhattacharjee, 'Globalising Economy, Localising Labour', *Economic and Political Weekly*, 14 October 2000.

3. Tirthankar Roy, 'Market-resurgence, Deregulation and Industrial Response: Indian Cotton Textiles in the 1990s', *Economic and Political Weekly*, Review of Industry, 1996.
4. NSSO, *Unorganized Manufacturing Enterprises in India: Its Size, Employment and Some Key Estimates. NSS Fifty-first Round*, Delhi, 1998; NSSO, *Tables with Notes on Survey of Unorganised Manufacture: Non-Directory Establishments and Own Account Enterprises. Part I: All India*, New Delhi, 1995; CSO, *Directory Manufacturing Establishments Survey 1989-90*, New Delhi, 1995.
5. For discussion and references, see A.C. Kulshreshtha and Gulab Singh, 'Gross Domestic Product and Employment in the Informal Sector of the Indian Economy', *Indian Journal of Labour Economics*, 42(2), 1999.

# Economic Reforms, Industrial Structure and Performance: The Case of Consumer Durable Goods Industry in India

K.V. RAMASWAMY

## 1. INTRODUCTION

What happens to industrial structure and performance when countries liberalize remains a keenly contested question. A large number of studies have focused on the impact of trade reforms on manufacturing productivity (Tybout, 1992). Similarly in India the attention is on the impact of industrial and trade reforms on efficiency and productivity (Srivastava, 2000 amongst others). One mechanism that has been traditionally emphasized is the impact of import competition on domestic firms in home markets. Import competition induced by trade reforms provides incentives for firms to improve their product quality and resource-efficiency. It is possible to have sectors, in the period of economic reforms, protected from imported goods but subject to competition from foreign firms undertaking local production and the free entry of domestic firms. This could be an outcome of perhaps the sequencing of reforms in context of a particular country. Economic theory does not provide clear predictions regarding policy changes and outcomes but only suggests several possibilities.[1] In this paper the impact of economic liberalization on the growth and structure of consumer durable goods industry in India has been taken up. Consumer durables is a generic term which is used for a set of industries that include traditional goods like wooden and steel furniture, bicycles, electric fans to technologically sophisticated goods like passenger cars, two-wheelers, colour TVs, washing machines, refrigerators and air-conditioners amongst others. Consumer durable (CD hereafter) industry has a special place in the Indian industrial structure in the sense that it has never been considered as a 'core' industry and belongs to the list of non-priority industries. The growth of CD was often considered to promote 'conspicuous consumption' or 'luxury goods' and adversely affect growth parameters like savings in the economy. With liberalization it turns out that CD industry is the fastest growing sector with high rates of investment bringing about structural transformation. The distinctive feature of these changes in the CD sector is that they were achieved in the absence of import competition. The performance of the CD sector, as evidenced in this study, is essentially an outcome of investment (domestic and foreign) and greater domestic competition in an environment of easier access to intermediate inputs, capital goods and foreign technology. This led to an enhancement of consumer welfare in terms of lower price and product variety. The impact of reforms is shown to be a virtuous circle of liberalization, investment, output growth and structural change.

The rest of this paper is organized as follows. In Section 2 the policy background for the study is presented. Section 3 contains a discussion of the data sources and period coverage. A brief discussion on consumption and consumer credit is also presented. Analysis of the growth performance of the CD industry is presented in Section 4. Section 5 presents the discussion of changing market structure and market shares. The impact of reforms on the entry, foreign investment, technology, prices and profitability is also discussed in Section 5. Section 6 contains a summary and the conclusions.

## 2. THE POLICY BACKGROUND

Even though policy history is common to all the sectors studied in this volume, it is useful to outline the policy changes that affected the CD industry.[2] This is pertinent because the timing and sequence of changes were not uniform across industry groups. Industrial licensing liberalization took place in stages and reforms permitting diversification were of greater importance to CD than to any other sector. Similarly, trade barriers like QR (Quantitative Restrictions) have continued to exist for CD while other industry groups have faced import competition fairly early. The foremost instrument of industrial policy was the industrial licensing for private entrepreneurs based on Industrial Regulation Act of 1956. An entrepreneur would have to apply and obtain a license from the government in order to set up a new unit, expand the existing capacity or manufacture a new product. There were barriers to entry on the larger firms with the Monopolies and Restrictive Trade Policies (MRTP) Act of 1970 and the Foreign Exchange Regulation Act (FERA) of 1973 being enacted. The MRTP restricted the capacity expansion of business houses with assets of more than Rs 200 million and those of 'dominant undertakings' with assets of more than Rs 10 million. The definition of dominance refers to a market share of more than 33 per cent. Any capacity expansion beyond 25 per cent of licensed capacity was regarded as 'substantial expansion', which required fresh government approval. This included any augmentation of production capacity resulting from modernization or renovation measures carried out by the company. Under the FERA regulation, foreign equity holdings had to be reduced to a maximum of 40 per cent. Agreements to import technology required prior government approval. The technology import policy prescribed royalty ceilings, duration of agreements, export obligations and sub-licensing conditions (Desai, 1985).[3] This adversely affected the content of technology made available to Indian companies.

Some initiatives to liberalize the licensing regime were taken in the years following the first oil price shock in 1973. In November 1975, 24 industries were taken out of the purview of IDR act. However, MRTP companies were denied entry into those 24 industries. New entrants were not permitted to import capital goods and work with foreign collaboration. Beginning in the year 1980, a list of industries were permitted to expand capacity without obtaining the 'substantial expansion' license. This list contained a large number of CD products. Another list of 25 industries was announced in March 1985 as delicensed, with similar conditions. Bicycles, pressure cookers and steel furniture were the only consumer durables that found a place in that list. MRTP companies were asked to take advantage of delicensing in electronic components and equipment subject to the condition of vertically integrated production, that is, no assembly type production. The fixed asset limit for firms under MRTP act was raised to Rs 1,000 million and the entry of MRTP/FERA companies was also permitted into a large number of industries by December 1985. The tyre industry was delicensed in the year 1989. In 1983, the more significant policy initiative of broadbanding was introduced, with the objective of facilitating full utilization of capacity. Under this policy, firms were permitted to shift from one product in the industry, say a light commercial vehicle to passenger car. The broadbanding facility was initially given to a set of 29 industry groups which was later extended to 35 industries in 1987. This set found a place for a large number of CD industries like automobiles, typewriters, electronics, air-conditioners and refrigerators. This policy of broadbanding is often considered as the beginning of industrial liberalization in India. However, it is important to recognize the restricted nature of this policy initiative. First, broadbanding scheme did not offer any scope to companies with fully utilized capacity, or potential producers of products reserved for small-scale industries. Second, companies could not diversify if the new product involved imported raw material or needed new foreign collaboration agreement. Third, if it involved substantial expansion then MRTP companies had to seek fresh approvals. Fourth, the union ministry of industry is reported to have stipulated an informal guideline of 15 per cent of company's investment for diversification purposes. In other words, broadbanding was allowed only in industries where the existing machinery and equipment could be used to make related products! Another offshoot of industrial licensing with the objective of preventing monopoly power was the establishment of firms with capacities below the minimum efficient economic scale of production. In recognition of this, government imposed minimum economic size condition on entry of firms in a number of industries. This condition encompassed those products reserved for small-scale industries like audio and videocassettes, colour TV receivers, tape recorders, etc. Entry into CD industries like passenger cars, tyres, two-wheelers and colour TV sets required setting up plants with minimum economic capacity. This condition was abolished with the new industrial policy of July 1991.

In response to the macro-economic crisis of July 1991 the government initiated industrial and trade reforms as part of the structural adjustment programme. The new industrial policy of 1991 abolished industrial licensing except in 18 industries. Motor cars, entertainment electronics (colour TV, tape recorders, CD players, etc.), and white goods (domestic refrigerators and washing machines, air-conditioners, etc.) were delicensed in later years. In 1993 the passenger car industry was delicensed followed by the white goods, and the entertainment electronics industry in 1996. Majority foreign equity participation was made automatic up to 51 per cent in 34 high-priority industry groups spread over 160 industries. Equity participation greater than 51 per cent was permitted on a case-by-case basis and the nature of production activity. Industries belonging to the CD sector were not included, as they were not considered as high-priority industries. Later the list was expanded to include CD industries. This enabled a large number of foreign companies to enter into joint ventures with Indian partners and set up 100 per cent subsidiaries. A number of consumer durables were under small-scale industry reservation like many electronic products, domestic appliances like iron, mixers, etc. These were removed from the reservation list enabling large domestic and foreign firms to enter those product lines. Simultaneously, the policy towards the foreign technology agreements was changed. Technology imports were permitted subject to the following conditions: lump sum payments should be below Rs 10 million, royalty on domestic sales below 5 per cent and on exports below 8 per cent, the total outgo over years from the date of agreement or 7 years from the date of agreement was below 8 per cent of sales. Non-priority industries like consumer good industries were subject to the foreign exchange cost balancing. That is a company's foreign exchange earnings must cover the cost of purchase of technology. Companies were free to employ foreign personnel. This was an important step as it improved the technology transfer process. Companies were also free to use foreign brand names and trademarks.

Trade policy was dominated by the concern to preserve foreign exchange through the system of import licensing. This implies licensing for all categories of imports except those listed in the Open General License (OGL) category. The relaxation of import restrictions began in the year 1977-8 (Pursell, 1992). However, all consumer goods continued in the banned list. A large number of capital goods, components and intermediate goods were shifted to the OGL list between 1976 and 1990. This change was made mainly to allow particular domestic industries to modernize and this shift was accompanied by reduced tariff rates. Most of the products shifted were those that were not produced domestically. Consequently, this would not have led to direct import competition for the incumbent firms (Pursell, 1992). In response to the balance of payment crisis of July 1991, the government announced trade reforms as part of the structural adjustment programme. This included the removal of import licensing and other non-tariff barriers on intermediate and capital goods. Reductions in tariffs on import of electrical machinery, non-electrical machinery, electronics and transport equipment in the range of 30 to 35 per cent indicated the onset of trade reforms. However, this process of trade reforms did not impact the CD industries as QRs continued to apply for all the consumer durables.[4] The quantitative restrictions on import of CD goods remained in place till April 2001. The impact of reforms on the CD industry was on the state of domestic competition in an environment of easier access to intermediate inputs, foreign investment and technology.

## 3. DATA SOURCES, PERIOD COVERAGE AND CONSUMPTION

In this study two main sources of data are used. First, data from the *Annual Survey of Industries* (ASI), published by the Central Statistical Organization (CSO) in New Delhi which includes all establishments with more than nine workers. I have selected 21 three-digit industry groups for analysis. These are listed with complete descriptions in the appendix. The analysis covers the period 1973-4 to 1997-8. The industrial classification system changed in the year 1973-4 and the data prior to 1973-4 is not comparable with those for post 1973-4. The second change in the classification system took place in 1989-90. Concordance tables have been used to match the three-digit industry groups within the CD sector. The industry groups between the two systems broadly match. However, a closer scrutiny of the data at the four-digit level revealed discrepancies with respect to many industry groups. The four-digit constituents of the three-digit group composition do not seem to match exactly. Consequently, growth rates have been estimated separately for the selected industries for the two sub-periods. The whole period has been divided into two sub-periods, namely, 1974-88 (pre-reform) and 1989-98 (post-reform). One could argue that this periodization is determined by the convenience of ASI data. This limitation persists but any alternative periodization would not change the conclusions and this also has the advantage of least adjustment to official data series. It is interesting to recall that the tyre industry was delicensed in June 1989, the first year of the post-reform period. One important adjustment needs to be mentioned. Till the year 1988-9, the industry group: motor vehicles and parts (374) includes data on heavy motor vehicles and coachwork (called 373 in the new series). In the new series, starting 1989-90, the industry group 374 refers to data on motorcars designed to transport less than 10 persons. The old series has been adjusted using the share of 373 and 374 in the new series. ASI reports data in current prices. So the choice of suitable deflators becomes important.[5] Gross Value Added (GVA) of the three-digit groups constituting the CD sector is deflated using industry specific Wholesale Price Index (WPI) with 1981-2 as the base year[6] in the price index of CD sector. This has been constructed as a chain index based on the WPI with 1981-2 as the base year.

The second set of data is based on the corporate database of the Centre for Monitoring Indian Economy (CMIE). CMIE collects data on all listed firms in the Bombay Stock Exchange. CMIE data has been used for the analysis of market shares, profitability, advertising expenditure, sources of finance, etc. Other sources are indicated at the relevant places in the text. It is useful to look at the evidence on changing consumption pattern of consumer durables in India before we move on to document the growth performance.

### *3.1. Consumption and Consumer Credit*

The per capita household consumption of consumer durables in India is expectedly very low and not comparable with those in industrialized societies.[7] It is the sheer size in terms of the number of consumer households that represents an enormous market potential for a variety of consumer durables. Successive household surveys of consumption report an increase in real per capita expenditures on non-food items and on consumer durables. According to the national accounts statistics data, the private final consumption expenditure in 1993-4, at current prices, on consumer durables was Rs 148.5 billion[8] which constituted little more than 3 per cent of the total consumption expenditure. The NSSO data estimates give the per capita household expenditure on durable items in 1993-4 as Rs 7.67 in rural India and Rs 15.17 in urban India. This constitutes only 2.7 per cent and 3.3 per cent of the total household spending in rural and urban sectors respectively. Given the relatively large size of rural India, this meagre expenditure translates into 60 per cent of national expenditure on consumer durables. The data on different fractile groups indicate that top 10 per cent of the population, spend more than 8 per cent of their per capita consumption expenditure on consumer durables. An alternative indicator of growth in consumption is suggested by data on stock penetration ratios of NCAER-MISH (1998) surveys.[9] The stock penetration ratio is defined as average number of household owning the durable goods in a population of 1,000 households. This source of data indicates that stock penetration ratios have continuously risen since 1985-6 across a variety of consumer durables like bicycles, cassette recorders, ceiling fans, two-wheelers, refrigerators, colour TVs and washing machines (see Table 24.3 in Natarajan, 1998). The penetration ratio in rural areas is still extremely low. Other surveys like ORG–MARG surveys show considerable ownership levels in households with income below Rs 50,000 per year in urban India. In this segment of the population 58 per cent have TVs, 10 to 25 per cent have a colour TV, refrigerator and two-wheeler, 50 per cent a pressure cooker, and 30 per cent LPG gas (Bijapurkar, 2000). Several factors would have contributed to this growth in consumption like the rapid growth of urban population and of households with middle incomes in urban areas. More important was the trend growth rate of 5.5 per cent and 7 per cent in total GDP and secondary and tertiary sectors respectively since 1980-1 and the emergence of consumer finance business. Consumer credit is an important determinant of demand for consumer durables in all countries. The post-reform period saw the emergence of consumer finance business as a rapid growth sector. It is reported that nearly 60 per cent of durable purchase were on hire-purchase basis.[10] All the major consumer durable companies were reported to have tied up with consumer finance companies to improve sales. In Table 1 the growth of bank credit to consumer durables since 1994 is presented. The estimates show that bank credit constitutes nearly 20 per cent of value added in the CD industry. This is admittedly an underestimation because commercial banks define the set of consumer durables eligible for bank finance rather narrowly.

TABLE 1. GROSS BANK CREDIT TO CD SECTOR*

| | *1995* | *1996* | *1997* | *1998* | *1999* | *2000* |
|---|---|---|---|---|---|---|
| CD credit | 1178 | 1476 | 2297 | 2527 | 3090 | 3855 |
| YOY% change | | 25.3 | 55.6 | 10.0 | 22.3 | 24.8 |
| % of GVA | 13.7 | 15.1 | 19.5 | 20.7 | NE | NE |

NOTES: * Outstanding as on at the end of March (Rs crore); NE: Not Estimated; GVA: Gross Value Added.

SOURCE: *Report on Trend and Progress of Banking in India*, RBI, various issues.

The estimate of value added in consumer durable industry is much broadly defined.[11] The estimates, shown in Table 1, suggest rapidly rising bank finance for the purchase of consumer durables. Banks finance constitute about 20 per cent of consumer durables value added in the years 1997 and 1998 as against 15 per cent in 1996.

## 4. GROWTH PERFORMANCE

Industrial structure in industrializing countries like India is largely determined by the size of the domestic market. A striking feature of recent industrial growth is the expansion of market size for consumer goods. In 1970-1 India produced about 50,000 cars, 2 million bicycles, 2 million electric fans and little less than 4 million automobile tyres. By the end of 1997-8 India produced about half a million cars, 30 million two-wheelers, 10 million bicycles, 8 million electric fans, and more than 30 million automobile tyres. Relaxation of supply constraints has been an important factor in this market expansion for consumer durables. The weight of the industry group consumer durables in the index of industrial production (IIP), a widely used measure of output growth, was just 2.5 in 1980-1. It had declined from the level of 5.7 in the year 1960-1. In the new series of IIP introduced in 1998, with the base year 1993-4, it regained prominence with a weight of 5.1. IIP shows that the CD sector is the best performing sector during the period 1994-5 to 1997-8 with average annual growth rate of 13.6 per cent. Over a longer period between 1991-2 and 1998-9 the CD sector registered a growth rate of 7.9 per cent when the entire manufacturing sector was growing at 5.8 per cent (Reserve Bank of India, 1998-9). Our estimates based on a more comprehensive database of ASI indicates that the CD sector

TABLE 2. GROWTH PERFORMANCE OF CD INDUSTRIES

| *Variable* | *Post-reform: 1989-90 to 1997-8* | *Pre-reform: 1973-4 to 1988-9* |
|---|---|---|
| Gross value added | 13.2 | 7.9 |
| Real wages | 4.8 | 6.0 |
| Capital formation (GFCF) | 16.1 | 12.1* |
| Employment | 3.9 | 2.3 |
| Labour productivity | 9.3 | 5.6 |

NOTE: * This is for the period 1979-80 to 1988-9.
SOURCE: Author estimate based on ASI summary results. Estimated as anti-logarithm of the regression coefficient minus one, with the equation Log y = a + bt.

has performed better in the post-reform period relative to a longer pre-reform period of 1973-4 to 1987-8. This is shown in Table 2. Notably, the trend growth rate of investment in CD industries was above 16 per cent, which was higher than that of the manufacturing sector (15 per cent). The virtuous circle of high investment, value added growth accompanied by higher labour productivity growth and improved real wage growth in high performance industry groups is shown in Table 3. The growth of output, investment, wages, employment and labour productivity of the entire set of industries in the post reform period is presented in Tables 4 and 5. The same information for the pre-reform is presented in Tables A1-A2 in the Appendix. The poor performers are bicycles and watches. Industrial policy reforms did not affect these two industries. Bicycle components and watches and clocks continue to be dominated by the small-scale sector (SSI) with product reservation. The trend of output growth of SSI segment has not been captured in this study. SSI also dominates wood products like boxes and furniture. Large fluctuations have been observed in capital formation in these groups. The performance of jewellery industry was not found to be exceptional in the post-reform period. This is again a traditional industry adding value to imported raw diamonds. Two-wheeler growth rate was found to be higher than that achieved during 1973-88 by about 3 percentage points. Foreign companies were permitted to enter two-wheeler industry through joint collaboration as far back as in 1982. This had led to the entry of Suzuki, Honda, and Yamaha into Indian industry through joint ventures with the Indian companies, namely, TVS, Hero and Escorts. The estimated high investment rate in the pre-reform period captures this entry activity. The other industry which needs to be mentioned is the tyre industry. Tyre industry was delicensed in June 1989. It had been facing high prices of its intermediate, rubber, in the international markets. Its growth rate remained moderate. Domestic electric appliances industry still has several of its products reserved for the SSI sector. Outsourcing was widely practiced (Ramaswamy, 1999). ASI data underestimates output in this industry.

In Table 6 the average annual percentage growth in sales, and the average export to sales and import to sales ratios for the period 1992-3 to 1997-8 has been presented. The data relates to a selected number of important CD industries. Except the tyre industry all other industries show a higher growth rate of sales than the manufacturing sector

TABLE 3. COMPARATIVE PERFORMANCE OF SELECTED CD INDUSTRIES

| *Industry* | *GVA* | *Real wage* | *GFCF* | *Lab. productivity* |
|---|---|---|---|---|
| *High Growth Groups* | | | | |
| Metal Furniture & Fixture | 40.9(6.1) | 32.0(7.8) | 43.0(14.4) | 27.6(6.6) |
| Metal Cutlery & Utensils | 14.1(-ve) | 5.6(1.1) | 43.0(-1.1) | 10.6(-ve) |
| Sewing & Washing Machines | 18.5(0.8) | 5.1(0.9) | 29.0(0.1) | 15.7(-ve) |
| TV Receivers & Audio Tapes | 17.4(11.6*) | 5.5(3.8) | 22.0(0.3) | 10.1(8.5) |
| Motor Cars & Parts | 17.6(6.7) | 5.1(5.7) | 21.0(5.1) | 15.8(3.9) |
| Motorcycles & Scooters | 16.5(13.4) | 4.6(11.8) | 14.0(39.0) | 12.4(4.7) |
| Photographic & Optical Goods | 31.1(7.9) | 15.0(0.11) | 46.0(19.6) | 18.3(8.6) |
| Jewellery | 27.5(11.4) | 22.0(2.3) | 19.0(98.0) | 11.7(10.5) |
| *Poor Performers* | | | | |
| Bicycles, Cycle Rickshaws | 5.8(16.3) | 2.7(3.7) | 6.8(2.7) | 2.2(11.8) |
| Watches & Clocks | (-ve)(13.5) | 3.9(8.9) | (-ve)6.1 | (**)7.4 |

NOTES: Figures in brackets refers to growth rates in the pre-reform period. Figures in the text are trend growth rates see the note to Table 2. Growth rates of GFCF are average of year over year changes for the period 1979-80 to 1988-9.

* Classification change; **Not estimated.

SOURCE: Authors estimates based on ASI.

TABLE 4. OUTPUT GROWTH AND INVESTMENT IN CD INDUSTRIES: 1989-90 TO 1997-8

| *NIC* | *Industry* | *GVA* | *Gross fixed capital formation* |
|---|---|---|---|
| 273 | Wooden Boxes, etc. | -7.7 | 18.7* |
| 276 | Wooden Furniture | -1.3 | 15.8* |
| 279 | Wood Products NEC | 16.3 | 16.7 |
| 310 | Tyres & Tubes | 10.1 | 16.2* |
| 342 | Metal Furniture & Fixture | 40.9 | 42.7 |
| 346 | Metal Cutlery & Utensils | 14.1 | 43.0 |
| 349 | Metal products NEC | 10.9 | 24.2 |
| 355 | Refrigerators & Air-conditioners | 4.6 | 7.49 |
| 358 | Office & Computing Machinery | 8.4 | 3.2* |
| 359 | Sewing & Washing Machine | 18.5 | 29.0 |
| 364 | Elec. Fans & Domestic Appliances | 11.9 | 16.4 |
| 365 | Radio & TV Apparatus | 3.5 | 4.2 |
| 366 | TV Receivers & Music Systems | 17.4 | 21.8 |
| 374 | Motor Cars & Parts | 17.6 | 21.3 |
| 375 | Motorcycles & Scooters | 16.5 | 14.0 |
| 376 | Bicycles & Cycle Rickshaws | 5.8 | 6.8 |
| 381 | Photographic & Optical Goods | 31.1 | 46.0 |
| 382 | Watches & Clocks | -6.6 | -10.9 |
| 383 | Jewellery & Related Articles | 27.5 | 18.8 |
| 386 | Musical Instruments | -7.8 | -5.3 |
| 389 | Miscellaneous Products | 17.2 | 29.1 |

NOTE: * Years with negative values and high positive values have been removed.

SOURCE: Author's estimates based on ASI summary results of various years.

TABLE 5. EMPLOYMENT, WAGES AND LABOUR PRODUCTIVITY IN CD INDUSTRIES: 1989-97

| *NIC* | *Industry* | *Real wage* | *Employment* | *Labour productivity* |
|---|---|---|---|---|
| 273 | Wooden Boxes, etc. | 1.2 | -0.3 | -7.4 |
| 276 | Wooden Furniture | 1.8 | 0.2 | -1.5 |
| 279 | Wood Products NEC | 8.5 | 3.8 | 12.5 |
| 310 | Tyres & Tubes | 4.0 | 4.1 | 6.0 |
| 342 | Metal Furniture & Fixture | 32.2 | 13.3 | 27.6 |
| 346 | Metal Cutlery & Utensils | 5.9 | 3.5 | 10.6 |
| 349 | Metal Products NEC | 8.8 | 8.3 | 2.6 |
| 355 | Refrigerators & Air-conditioners | -1.7 | -0.4 | 5.0 |
| 358 | Office & Computing Machinery | -3.9 | -5.7 | 14.1 |
| 359 | Sewing & Washing Machine | 5.2 | 2.8 | 15.7 |
| 364 | Elec. Fans & Domestic Appliances | 0.8 | 1.8 | 10.1 |
| 365 | Radio & TV Apparatus | 2.2 | 1.5 | 2.0 |
| 366 | TV Receivers & Music Systems | 5.5 | 1.6 | 15.8 |
| 374 | Motor Cars & Parts | 5.1 | 3.6 | 14.0 |
| 375 | Motorcycles & Scooters | 4.6 | 4.1 | 12.4 |
| 376 | Bicycles & Cycle Rickshaws | 2.7 | 4.5 | 1.3 |
| 381 | Photographic & Optical Goods | 15.2 | 12.8 | 18.3 |
| 382 | Watches & Clocks | 3.9 | 0.3 | -6.9 |
| 383 | Jewellery & Related Articles | 22.2 | 15.8 | 11.7 |
| 386 | Musical Instruments | -4.7 | -4.9 | -2.9 |
| 389 | Micellaneous Products | 0.3 | 4.1 | 13.1 |

NOTE: *In these industries years (1996-7) with negative GFCF were deleted.
SOURCE: ASI various years.

TABLE 6. GROWTH IN SALES, EXPORTS AND IMPORTS IN CD INDUSTRIES

| | *1992-8* | | |
|---|---|---|---|
| *Industry groups* | *Sales** | *Export/Sales*** | *Import/Sales*** |
| Tyres & Tubes | 9.4 | 9.7 | 11.0 |
| Domestic Electrical Appliances | 14.6 | 5.8 | 9.2 |
| Air-conditioners & Refrigerators | 18.1 | 2.3 | 8.2 |
| Consumer Electronics | 16.6 | 4.6 | 14.1 |
| Passenger Cars & MUV | 18.9 | 5.1 | 16.8 |
| Two- & Three-wheelers | 21.1 | 4.9 | 8.1 |
| Manufacturing | 14.7 | 8.0 | 15.8 |

NOTES: * Average annual percentage growth.
** Average of annual percentage.
SOURCE: CMIE Corporate data base.

average. Their export intensities were also found to be below the manufacturing average indicating the domestic market orientation of the CD industry. I have also examined the growth of import intensities

TABLE 7. SOURCES OF FUNDS IN THE CD INDUSTRY: 1992-8

*(average percentage for 1992-8)*

| *Industry groups* | *Bank borrowings** | *Internal sources** | *Raised from in capital markets** |
|---|---|---|---|
| Tyres & Tubes | -ve | 46.7 | 18.8 |
| Domestic Electrical Appliances | 23 | 15.6 | 26.9 |
| Air-conditioners & Refrigerators | 7.3 | 34.0 | 18.2 |
| Consumer Electronics | 11.1 | 29.9 | 32.6 |
| Passenger Cars & MUV | -vc | 31.5 | 19.1 |
| Two- & Three-wheelers | 1.1 | 61.1 | 6.9 |
| Manufacturing | 13.6 | 27.5 | 23.3 |

NOTE: * As percentage of total funds.
SOURCE: CMIE data.

over the period 1992 to 1998 and found no alarming increase. They only reflect greater access to intermediate inputs essential to quality outputs. The CD industry also seems to have raised more finance through internal sources compared to the average for the manufacturing companies as shown in Table 7.[12] Now the impact of reforms on entry of firms, market shares and restructuring process in the CD industry will be duly examined.

## 5. MARKET STRUCTURE, ENTRY AND RESTRUCTURING OF THE CD INDUSTRIES

### *5.1 Market Structure and Market Shares*

Market structure determines the conduct of firms and in turn determines performance. Simultaneous relationship between structure and performance variables is well recognized. In India the industrial policy brought into existence a particular structure and further failed to provide any scope or incentives for change in the structure. Oligopoly and duopoly characterized the market structure of CD industries. Oligopoly is the feature of many CD industries even today with a difference. The threat of potential entry and strategic competition determines behaviour (hence outcomes) not captive markets. Reliable market share data has always been and continues to be scarce. Market share is more subject to errors in industries where large unorganized sector co-exists with the organized segment. For example, in industries like air-conditioners, steel furniture, air-coolers and domestic electrical appliances. Relevant statistics have been extracted from three previous studies, namely, Desai (1984), Mani (1993) and Khatkhate (1992). They are presented in Tables 8 and 9 respectively. The measures of concentration were high in motor vehicles and auto tyres and low in bicycles and refrigerators. Few firms dominated in most of the CD industries. Hindustan Motors, Premier Automobiles in passenger cars, Enfield India and Escorts in motor cycles, Bajaj Auto in scooters, Godrej & Boyce in steel furniture, Singer in sewing machines, Kelvinator of India and Godrej & Boyce in refrigerators. The liberalization of industrial policies changed the entry conditions.

In response existing firms expanded capacities, domestic and foreign firms created new capacities. Competition for market share led to a decline in H-H indices of concentration. They are shown in Table 10. Concentration indices are much lower for the motor vehicle industries that prevailed in the pre-reform period indicated in Tables 8 and 9. Exceptions do exist. Three industries show a tendency to rise. These are motorcycles, bicycles and passenger cars. The first two are due to the dominance of Hero Cycles and Hero-Honda. The trend in passenger cars reflects the dominance of MUL (Maruti Udyog Limited), the manufacturers of Maruti-Suzuki cars.

TABLE 8. H-H INDEX IN SELECTED CD INDUSTRIES IN 1978-9

| *Industry* | *Herfindahl index* |
|---|---|
| Bicycles | 0.13 |
| Cars | 0.71 |
| Scooters | 0.30 |
| Motorcycles | 0.36 |
| Refrigerators | 0.13 |

NOTE: H-H Index: Herfindahl-Hirschman Index.
SOURCE: Desai (1984).

TABLE 9. FOUR FIRM CONCENTRATION RATIOS IN SELECTED CD INDUSTRIES

| *Industry* | *1983-4* | *1988-9* |
|---|---|---|
| Cars | 100 | 100 |
| Motorcycles | 97 | 93 |
| Tyres | 61 | 66 |
| Mopeds | 97* | 86** |
| Scooters | 96* | 96** |
| Electric Fans | 81* | 69** |

NOTES: *1984-5; **1987-8.
SOURCE: Mani (1993) and Khatkhate (1992) based on CMIE data.

TABLE 10. H-H INDEX OF CONCENTRATION FOR CD INDUSTRIES

| *Industry* | *1993-4* | *1998-9* |
|---|---|---|
| Tyres | 0.109 | 0.132 |
| Air-coolers | 0.673 | 0.673 |
| Air-conditioners | 0.203 | 0.165 |
| Electrical Fans | 0.143 | 0.158 |
| Washing Machines | 0.361 | 0.307 |
| Refrigerators | 0.349 | 0.235 |
| Audio Systems | 0.355 | 0.359 |
| Scooters | 0.367 | 0.322 |
| Three-wheelers | 0.850 | 0.522 |
| Mopeds | 0.232 | 0.301 |
| Motorcycles | 0.252 | 0.261 |
| Bicycles | 0.233 | 0.299 |
| Passenger Cars | 0.475 | 0.542 |
| Watches & Clocks | 0.295 | 0.297 |

NOTE: HH Index: Herfindahl-Hirschman.
SOURCE: CMIE Corporate data.

TABLE 11. NUMBER OF COMPANIES IN CD INDUSTRIES

| *Industry group* | *1992-3* | *1995-6* | *1998-9* |
|---|---|---|---|
| Tyres & Tubes | 19 | 23 | 20 |
| Domestic Electrical Appliances | 30 | 40 | 19 |
| Air-conditioners & Refrigerators | 8 | 14 | 12 |
| Consumer Electronics | 22 | 38 | 25 |
| Passsenger Cars & MUV | 5 | 7 | 6 |
| Two- & Three-wheelers | 12 | 13 | 12 |
| Manufacturing | 2,506 | 4,422 | 3,651 |

SOURCE: CMIE data base.

This section will be concluded by presenting two important pieces of evidence on entry. First, Table 11 indicates an increase in the number of companies in 6 important industries. Second, I estimate the net entry of factories between 1992 and 1998 in the factory sector.[13] They are shown in Table 12. Net entry is positive in most of the 21 industries except for 6 of them.[14] Net entry reflects the plant turnover process in the CD industries. Many studies draw attention to the importance of entry and exit of plants in the industrial restructuring process.[15] It is the entry of plants with new technology and the gradual substitution of old plants by efficient plants that lead to productivity growth. Now the other important feature of the CD industry during the reform period will be studied in the following sub-section.

*5.2 Factory Size, Foreign Investment and Technology*

A well-known trend in the Indian industries is the observed continuous decline in establishment (factory) size. Here, workers per factory measure factory size. Several factors could contribute to this phenomenon. Entry of factories below minimum efficient scale could be an important factor. Fragmentation of factories to take advantage of government incentives is another. Some weak evidence can be found to indicate that this tendency was being reversed in the case of CD industries. In the Appendix: Table A3, estimates of factory size in 8 selected CD industries are shown for the post-reform period, that is 1989 to 1998. In most of the industries factory size is shown to be rising. The exception seem to be the year 1997-8, a year of low employment growth. If this finding is accepted then it indicates another positive impact of industrial reforms.

Firms in the CD industry actively sought foreign technology and finance. During this period FDI (Foreign Direct Investment) approvals amounted to more than Rs 112 billion in transportation equipment, Rs 95 billion in electrical equipment and electronics, and Rs 9.8 bil-

TABLE 12. NET ENTRY OF FACTORIES IN CD INDUSTRY: 1990-8

| *Industry code* | *No. of factories 1989-90* | *No. of factories 1997-8* | *Increase/Decrease in no. of factories* |
|---|---|---|---|
| 310 | 361 | 551 | 190 |
| 342 | 383 | 355 | -28 |
| 346 | 1,348 | 1,473 | 125 |
| 349 | 400 | 608 | 208 |
| 355 | 155 | 230 | 75 |
| 358 | 60 | 73 | 13 |
| 359 | 1,091 | 1,135 | 44 |
| 364 | 630 | 610 | -20 |
| 365 | 283 | 294 | 11 |
| 366 | 529 | 470 | -59 |
| 374 | 1,008 | 1,229 | 221 |
| 375 | 550 | 534 | -16 |
| 376 | 693 | 654 | -39 |
| 381 | 107 | 117 | 10 |
| 382 | 219 | 200 | -19 |
| 383 | 270 | 449 | 179 |
| 386 | 11 | 13 | 2 |
| Total | 9,081 | 10,150 | 1,069 |

SOURCE: Author's estimates based on ASI summary results. For Industry Description see Appendix A.

lion in commercial, office and household equipment. Data on number of approvals is presented in Table 13. This sufficiently indicates the inflow of foreign investment and technology to CD industries. FDI in more narrowly defined industry categories is not available. The question whether greater access to foreign technology in the reform period led to less local R&D effort is not yet a settled issue. The evidence on Research and Development intensities (R&D/Sales ratio), presented in Table 14, suggests that the average ratio of CD industries is more than that of the total manufacturing companies. Other important Indian companies are reported to have increased their R&D expenditures manifold during the reform years. Individual company data has not been tracked in this study. An environment of strategic competition unleashed by the reform is unlikely to slow down the technological efforts of companies in the CD industries.[16]

*5.3 Prices and Profitability*

The entry of foreign firms, expansion and diversification of Indian firms brought new technologies, new differentiated products with foreign brand names, and intensified price and non-price competition in most of the CD industries. New entrants penetrated the markets of incumbents through aggressive pricing, advertising, investment in distribution networks and other promotional strategies. Incumbents responded to this threat in order to defend their market shares.[17] Consequently, the relative prices of consumer durables declined in the post-reform period. This process was facilitated by the reform and rationalization of indirect taxation in the successive budgets by the Government of India. The tax and trade reforms brought down the excise tax rates and the customs duties on components used in the CD industry. In Figure 1 the trend in the relative prices of consumer durables is shown for the period 1993 to 1998. This represents the welfare enhancing impact of economic reforms in India. The competition for markets resulted in the introduction of new models and brands of consumer durables positively affecting consumer welfare through product variety.[18] Advertisement expenditures have grown rapidly in the period of liberalization. These have grown faster in the CD industry relative to the growth rate for all manufacturing companies, a fact which is supported by the estimates presented in Table 15. One is inclined to the view that advertising can be pro-competitive if it contributes to the awareness about the entrant's

TABLE 13. FOREIGN COLLABORATION CASES APPROVED BETWEEN 1991 TO 1998

| *Industry* | *No. of approvals* | | | *Percentage of total* |
|---|---|---|---|---|
| | *Total* | *Technical* | *Financial* | *value of FDI approved* |
| Transportation (includes Automobile) | 895 | 433 | 462 | 6.2 |
| Electrical Equipment (Computer software and Electronics) | 2,490 | 987 | 1,503 | 5.2 |
| Commercial, Office and Household Equipment | 70 | 27 | 43 | 0.5 |

SOURCE: Annual Report of Ministry of Industry, GOI, 1998-9, pp. 28-9.

TABLE 14. R&D INTENSITIES OF SAMPLE ENTERPRISE IN SELECTED INDUSTRY GROUPS. 1992-8

*(average percentage of sales)*

| *Industry* | *Local firms* | *MNE affiliates* | *Total* |
|---|---|---|---|
| Metal and Metal Products | 0.47 | 0.50 | 0.47 |
| Automobiles | 1.33 | 0.79 | 1.12 |
| Electrical Equipment | 1.10 | 0.90 | 1.08 |
| All Firms in the sample | 0.85 | 0.82 | 0.85 |

SOURCE: Kumar and Agrawal (2000).

TABLE 15. GROWTH OF ADVERTISING EXPENSES 1992-8

| *Industry groups* | *Average growth rate* | *Average advertisement intensity** |
|---|---|---|
| Tyres & Tubes | 12.4 | 1.3 |
| Domestic Electrical Appliances | 17.7 | 2.1 |
| Air-conditioners & Refrigerators | 60.8 | 3.4 |
| Consumer Electronics | 30.0 | 3.6 |
| Passenger Cars & MUV | 68.7 | 0.5 |
| Two- & Three-wheelers | 32.3 | 2.0 |
| All Manufacturing Firms | 24.8 | 0.7 |

NOTE: * Advertisement to sales ratio.
SOURCE: CMIE data base.

products and that advertising had this particular effect in the post-reform period.

What has been the trend in profitability during the post-reform period? A profitability comparison between the pre-reform and the post-reform period is fraught with difficulties.[19] Here the profitability ratios of companies in the selected CD industries using the CMIE database are considered only. They are shown in Table 16. Except in

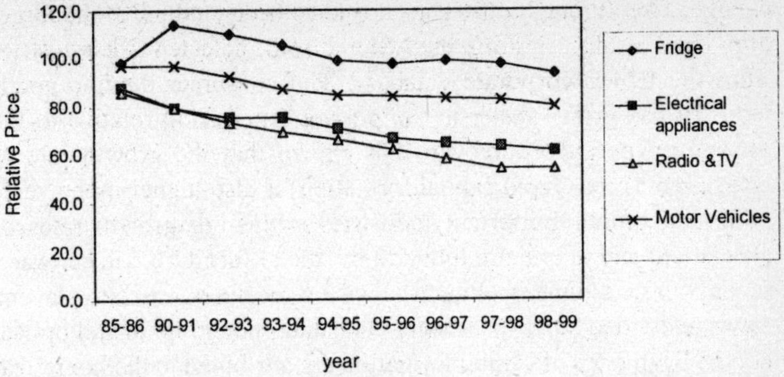


FIGURE 1. RELATIVE PRICE CHANGE IN CONSUMER DURABLES

TABLE 16. PROFITABILITY IN THE CD INDUSTRY: 1993-8

| *Industry groups* | *1992-3* | *1993-4* | *1994-5* | *1995-6* | *1996-7* | *1997-8* | *1998-9* |
|---|---|---|---|---|---|---|---|
| Tyres & Tubes | 9.8 | 8.4 | 8.3 | 9.8 | 8.8 | 10.8 | 11.0 |
| Domestic Electrical Appliances | 10.5 | 11.6 | 11.0 | 9.9 | 10.5 | 9.9 | 10.0 |
| Air-conditioners & Refrigerators | 5.3 | 8.2 | 5.6 | 5.9 | 4.3 | 4.2 | 6.6 |
| Consumer Electronics | 9.6 | 9.4 | 10.5 | 8.3 | 8.3 | 9.2 | 9.6 |
| Passenger Cars & MUV | 6.6 | 10.1 | 11.6 | 12.8 | 11.1 | 14.3 | 11.0 |
| Two- & Three-wheelers | 11.1 | 15.4 | 16.5 | 15.8 | 16.8 | 16.0 | 15.5 |
| Manufacturing | 12.1 | 12.5 | 13.0 | 13.3 | 12.5 | 12.2 | 11.7 |

NOTE: *PBDIT* as a percentage of sales.
SOURCE: CMIE data.

the industry group of two- and three-wheelers, average profitability do not seem to show a rising trend. They are lower than the average profitability of all manufacturing firms.

## 6. SUMMARY AND CONCLUSION

In this study the impact of economic reforms on the growth and structure of the consumer durable (CD) goods industry in India has been investigated. The objective of economic policy change is to create greater domestic competition in an environment of easier access to intermediate inputs, capital goods and foreign technology. This study shows that the outcome of reforms is a virtuous circle of liberalization of entry, investment, output growth and structural change. Twenty-one industry groups have been studied, at the three-digit level, using data from the ASI and some selected CD industries from the CMIE corporate database. The consumer durable goods industry has grown faster in the post-reform period relative to the pre-reform period. Moreover, it is shown that the growth rate of investment (gross fixed capital formation) is also higher in the years of liberalization. Important industries with high growth rates of investment and output are found to be: metal furniture, kitchenware, sewing and washing machines, television receivers, cassette players, audio and video tapes, passenger cars, and photographic and optical goods. High rates of capital formation are attributed to the liberalization of licensing, entry and expansion conditions for domestic and foreign firms. Labour productivity, employment and wages are found to have risen faster in a number of CD industries in the reform years. A significant finding is the changing market structure of CD industries in favour of lower levels of concentration. This restructuring of the CD industries is a significant positive outcome of industrial policy reforms. Evidence presented clearly shows that the real prices of consumer durables, relative to all manufacturing goods, have declined in the post-reform period. This was an outcome of the creation of competitive conditions in the CD industries. On this basis, it can be concluded that the impact of economic reforms was welfare enhancing in terms of lower consumer prices and product variety.

Economic reforms have been successful in forcing firms in the CD industry to restructure themselves to operate in competitive environment. Firms have responded by making investment in economically efficient capacities and technological upgradation. The challenge seems to be in the ability to harness this change into productivity growth. On this there is lack of evidence. This kind of restructuring which responds to changes in markets, technology and environmental regulations is widely accepted to be a continuous process. The real test of firm capabilities, in the CD industry, to restructure in the face of such changes is beginning to unfold. For example, removal of quantitative restrictions in April 2001, and the necessity to introduce technologies to meet emission standards. In this sense, rapid changes are likely to be seen in the growth and structure of the CD industry.

# APPENDIX A: TABLES

TABLE A1. OUTPUT AND INVESTMENT GROWTH IN CD INDUSTRY: 1973-88

| *NIC-70* | *Industry* | *Gross value added* | *Gross fixed capital formation*** |
|---|---|---|---|
| 272 | Wooden Boxes | 2.5 | ** |
| 276 | Wooden Furniture | 8.5 | ** |
| 279 | Wood & Bamboo Products | 31.8 | ** |
| 300 | Tyre & Tube Industries | 9.1 | Negative |
| 342 | Furniture & Fixture | 6.1 | 14.4 |
| 345 | Metal Utensils & Cutlery | 0.0 | -1.1 |
| 349 | Metal Products | 3.8 | 26.0 |
| 355 | Refrigerators & Air-conditioners | 6.1 | 31.0 |
| 358 | Office Computing & A/c Machinery | 4.8 | 53.0 |
| 359 | Sewing Machines/Washing Machines | 0.8 | 14.1 |
| 363 | Electrical Lamps, etc. | 11.6 | 16.5 |
| 364 | Radio, TV, Music Apparatus | 11.6 | 30.0 |
| 374 | Motor Vehicles & Parts | 6.7 | 5.1 |
| 375 | Motorcycles, Scooters | 13.6 | 39.0 |
| 376 | Bicycles, Cycle-Rickshaws | 16.4 | 20.00 |
| 381 | Photographic & Optical Goods | 8.0 | 19.3 |
| 382 | Watches & Clocks | 13.5 | 54.0 |
| 383 | Jewellery, etc. | 11.4 | 98.0 |
| 386 | Musical Instruments | -17.9 | -14.5 |
| 389 | Miscellaneous n.e.c. | 3.7 | 15.0 |

NOTE: GFCF is average of YOY change.
SOURCE: Author's estimate based on ASI.

TABLE A2. GROWTH OF CONSUMER DURABLE GOODS INDUSTRY: 1973-88

| *NIC-70* | *Industry* | *Real wage* | *Employment* | *Labour productivity* |
|---|---|---|---|---|
| 272 | Wooden Boxes | 0.2 | -4.0 | 6.5 |
| 276 | Wooden Furniture | 1.2 | -3.4 | 11.9 |
| 279 | Wood & Bamboo Products | 16.7 | 12.1 | 19.7 |
| 300 | Tyre & Tube Industries | 5.5 | 3.8 | 5.3 |
| 342 | Furniture & Fixture | 7.8 | -0.5 | 6.6 |
| 345 | Metal Utensils & Cutlery | 1.1 | 1.1 | -1.1 |
| 349 | Metal Products | 3.3 | -0.1 | 3.9 |
| 355 | Refrigerators & Air-conditioners | 3.6 | 0.2 | 5.9 |
| 358 | Office Computing & A/c Machinery | 0.8 | -1.0 | 5.9 |
| 359 | Sewing Machines/Washing Machines | 0.9 | 3.2 | -2.4 |
| 363 | Electrical Lamps, etc. | 3.9 | 1.7 | 9.9 |
| 364 | Radio,TV, Music Apparatus | 3.8 | 3.1 | 8.5 |
| 374 | Motor Vehicles & Parts | 5.7 | 2.9 | 3.8 |
| 375 | Motorcycles, Scooters | 11.8 | 8.7 | 4.8 |

(Contd.)

TABLE A2 (Contd.)

| *NIC-70* | *Industry* | *Real wage* | *Employment* | *Labour productivity* |
|---|---|---|---|---|
| 376 | Bicycles, Cycle Rickshaws | 3.7 | 2.7 | 13.7 |
| 381 | Photographic & Optical Goods | 0.1 | -0.7 | 8.6 |
| 382 | Watches & Clocks | 8.9 | 6.1 | 7.4 |
| 383 | Jewellery, etc. | 2.3 | 0.9 | 10.5 |
| 386 | Musical Instruments | -4.4 | -2.8 | -15.1 |
| 389 | Miscellaneous n.e.c. | 4.0 | -0.5 | 4.2 |

NOTE: GFCF is average of year over year change.
SOURCE: Author's estimate based on ASI summary results.

TABLE A3: FACTORY SIZE IN CD INDUSTRIES: 1990-8

| *Year* | *376* | *375* | *374* | *381* | *365* | *364* | 342 | 383 | 349 | 310 |
|---|---|---|---|---|---|---|---|---|---|---|
| 89-90 | 42.68 | 80.10 | 69.52 | 18.51 | 109.95 | 22.93 | 12.70 | 37.57 | 26.6 | 104.77 |
| 90-1 | 46.89 | 97.21 | 64.85 | 17.13 | 104.95 | 27.21 | 13.22 | 46.57 | 26.7 | 106.77 |
| 91-2 | 42.23 | 80.29 | 67.34 | 17.89 | 114.99 | 26.21 | 12.90 | 62.24 | 26.9 | 96.04 |
| 92-3 | 41.23 | 85.59 | 67.24 | 19.99 | 110.66 | 26.31 | 25.97 | 52.10 | 27.5 | 101.46 |
| 93-4 | 41.93 | 80.56 | 66.00 | 22.41 | 103.98 | 32.39 | 24.87 | 72.48 | 26.6 | 97.77 |
| 94-5 | 48.41 | 87.64 | 70.93 | 20.28 | 108.06 | 30.25 | 23.97 | 66.60 | 29.6 | 96.92 |
| 95-6 | 54.35 | 110.36 | 73.46 | 28.18 | 122.63 | 34.28 | 28.56 | 68.24 | 30.4 | 90.14 |
| 96-7 | 82.42 | 111.63 | 73.73 | 27.80 | 125.90 | 29.26 | 27.47 | 78.33 | 29.0 | 102.74 |
| 97-8 | 51.99 | 102.85 | 74.47 | 49.23 | 111.59 | 24.09 | 36.88 | 68.36 | 30.9 | 83.57 |

SOURCE: Author's estimate based on ASI data.

## NOTES

1. See Ho Aswicahyono et al., 'What Happens to Industrial Structure when Countries Liberalize? Indonesia since the Mid 1980s', *The Journal of Development Studies*, Vol. 32, No. 2, 1996, for a broader discussion.
2. In this section I have traced the policy reforms up to the year 1998-9. This description relies heavily on the details available in the document 'The Liberalization Process' published by the Centre for Monitoring Indian Economy, Mumbai (1986). *The Economic Survey of Government of India* published annually documents the reform measures. *Annual Report* of the ministry of industry is another important source.
3. Ashok Desai, 'Indigenous and Foreign Determinants of Technological Change in Indian Industry', *Economic and Political Weekly*, Vol. 20, Nos. 45 to 47, 1985, pp. 2081-94, provides an analysis of impact of the restrictive nature of technology policy on technology development in India.
4. Imports of some durables under the restricted list are permitted under licensing since 1992. But this will not affect my conclusions.
5. Bishwanath Goldar, 'Productivity Growth in Indian Manufacturing in the 1980s and 1990s', 20 November 2000, Institute of Economic Growth, discusses this problem in the context of productivity growth measurement.
6. The deflator for the CD industry is a weighted average of inflation rates of the constituent three-digit industry groups with value added shares as weights. See contribution by Shuji Uchikawa in this volume.
7. The consumer durable goods are distinct from other consumer goods in important ways. They embody modern production technology, their purchase involves bulky expenditure, and utility is spread over a period with features of indivisibility, that is members of the household together benefit from a unit in possession. Their use requires certain preconditions like power and fuel.
8. This is based on figures reported in K. Sundaram and Suresh Tendulkar, 'NAS-NSS Estimates of Private Consumption for Poverty Estimation: A Disaggregated Comparison for 1993-94', *Economic and Political Weekly*, Vol. 36, No. 2, 2001, pp. 119-29.
9. National Council of Applied Economic Research (NCAER) in Delhi conducts periodic surveys of household consumption of selected manufactured goods. These are called Market Information Survey of Households (MISH). The latest survey has 1995-6 as the reference year. We abstain from a critical scrutiny of the methodology of these surveys as many details are not in the public domain.
10. This is stated in the industry reports available online at www.indiainfoline.com. Non Bank Financial Companies (NBFCs) play an important role in this business.
11. In my estimation of Gross Value Added, for this purpose, I exclude only wood products, tyres, photographic goods, jewellery and watches.
12. Many companies in the CD industry are foreign controlled companies with access to deep pockets. More research is called for in this area of investment financing.
13. Here factories are synonymous with establishments or plants in the language of industrial economics literature.
14. Note that this is different from gross entry. Gross entry measures the number of new plants in the market over a period. Net entry simply counts the change in

the number of plants over a given period. It is hard to get data on gross entry. The use of net entry here is sufficient to indicate the extent of plant turnover in the CD industries.

15. See the studies in Mark Roberts and James Tybout, 'Industrial Evolution in Developing Countries: Micro Patterns of Turnover, Productivity and Market Structure', New York: Oxford University Press, 1996.
16. For some evidence see the report on Videocon and Moser Baer in *Business World*, 28 August 2000.
17. Reports in the business magazines are full of stories about how Maruti in cars, Bajaj in two-wheelers and HMT in watches are restructuring to face threat of entrants. I have not attempted to document such processes in this study.
18. The business magazines have frequently highlighted the fall in the prices of colour TVs and refrigerators in the post-reform period. An example cited is the availability of electrolux 165-litre model for Rs 7,000 in 1998, the same price was in 1991.
19. Analysis of financial statements of diversified firms poses problem of comparability. Apart from this difficulty the major problem is one of distinguishing the effects of recessionary conditions and competitive outcomes. A rigorous analysis of profitability trends needs a separate paper.

## REFERENCES

Ahluwalia, Isher J., *Industrial Growth in India: Stagnation Since the Mid-Sixties*, New Delhi: Oxford University Press, 1985.

Acwicahyono, H., Kelly Bird and Hal Hill, 'What Happens to Industrial Structure When Countries Liberalize? Indonesia since the Mid-1980s', *The Journal of Development Studies*, Vol. 32, No. 3, 1996, pp. 340-63.

Bijapurkar, Rama, 'The Marketing of India', *The Economic Times*, 18 September 2000.

Desai, Ashok, *Market Structure and Technology: Their Interdependence in Indian Industry*, New Delhi, Working Paper 13, Indian Council for Research in International Economic Relations (ICRIER).

——'Indigenous and Foreign Determinants of Technological Change in Indian Industry', *Economic and Political Weekly*, Vol. 20, Nos. 45 to 47, 1985, pp. 2081-94.

Goldar, Bishwanath, 'Productivity Growth in Indian Manufacturing in the 1980s and 1990s', 20 November, Institute of Economic Growth, New Delhi.

Jacobsson, Staffan, 'Government Policy and Performance in the Indian Engineering Industry', *Research Policy*, Vol. 20, 1991, pp. 45-56.

Khatkhate, Deena, 'The Regulatory Impediments to the Private Industrial Sector Development in Asia: A Comparative Study', *World Bank Discussion Paper*, Washington, DC, No. 177, 1992.

Kumar, Nagesh and Aradhna Agrawal, 'Liberalisation, Outward Orientation and In-house R&D Activity of Multi National and Local Firms: A Quantitative Exploration for Indian Manufacturing', paper presented at the conference to honour Professor K.L. Krishna at Delhi School of Economics, 20-3 December 2000, New Delhi.

Ministry of Industry, *Annual Report 1989-90*, Government of India, New Delhi.

Mani, Sunil, 'Changing Structure of India's Manufacturing Sector', paper presented at the Seminar on New Economic Policy, 19-21 August 1993 at the Indian Institute of Management, Calcutta.

Natarajan, I., *India Market Demographics Report 1998*, National Council of Applied Economic Research, 1998, New Delhi.

Pursell, G., 'Trade Policies in India', in D. Salvatore (ed.), *National Trade Policies*, New York: Greenwood Press, 1992.

Ramaswamy, K.V., 'The Search for Flexibility in Indian Manufacturing: New Evidence on Outsourcing Activities', *Economic and Political Weekly*, Vol. 34, No. 6, 1999, pp. 363-8.

Reserve Bank of India (RBI), *Report on Currency and Finance 1998-99*, Mumbai: Reserve Bank of India.

Roberts, Mark and James Tybout, *Industrial Evolution in Developing Countries: Micro Patterns of Turnover, Productivity and Market Structure*, New York: Oxford University Press, 1996.

Srivastava, Vivek, 'The Impact of India's Economic Reforms on Industrial Productivity, Efficiency and Competitiveness', report of a project sponsored by the Industrial Development Bank of India, National Council of Applied Economic Research, 2000, New Delhi.

Sundaram, K. and Suresh Tendulkar, 'NAS-NSS Estimates of Private Consumption for Poverty Estimation: A Disaggregated Comparison for 1993-94'. *Economic and Political Weekly*, Vol. 36, No. 2, 2001, pp. 119-29.

Tybout, James, 'Linking Trade and Productivity: New Research Directions', *The World Bank Research Observer*, Vol. 6, No. 2, 1992, pp. 189-211.

# Appendix

## CAPITAL GOOD INDUSTRY GROUP

| *NIC Code (1987)* | *Industries* |
|---|---|
| 340 | Manufacture of fabricated structural metal products |
| 341 | Manufacture of fabricated metal products not elsewhere classified |
| 343 | Manufacture of hand tools, weights and measures and general hardware |
| 350 | Manufacture of agricultural machinery and equipment and parts thereof |
| 351 | Manufacture of machinery and equipment used by construction and mining industries |
| 352 | Manufacture of prime movers, boilers, steam generating plants and nuclear reactors |
| 353 | Manufacture of industrial machinery for food and textile industries (including bottling and filling machinery) |
| 354 | Manufacture of industrial machinery for other than food and textile industries |
| 356 | Manufacture of general purpose non-electrical machinery/equipment, their components and accessories n.e.c. |
| 357 | Manufacture of machine tools, their parts and accessories |
| 360 | Manufacture of electrical industrial machinery, apparatus and parts thereof |
| 367 | Manufacture of computers and computer based systems |
| 369 | Manufacture of radiographic X-ray apparatus, X-ray tubes and parts and manufacture of electrical equipment n.e.c. |
| 370 | Ship and boat building |
| 371 | Manufacture of locomotives and parts |
| 372 | Manufacture of railway/tramway wagons and coaches and other railroad equipment n.e.c. |
| 373 | Manufacture of heavy motor vehicles; coach work |
| 377 | Manufacture of aircraft, spacecraft and their parts |
| 378 | Manufacture of bullock-carts, push-carts and hand-carts, etc. |
| 379 | Manufacture of transport equipment and parts not elsewhere classified |
| 380 | Manufacture of medical, surgical, scientific and measuring equipment except optical equipment |
| 388 | Manufacture of items based on solar energy like solar cells, cookers, air and water heating systems and other related items |
| 39 | Repair of capital goods |

# INTERMEDIATE GOODS INDUSTRY GROUP

| *NIC Code (1987)* | *Industries* |
|---|---|
| 230 | Cotton ginning, cleaning and baling |
| 231 | Cotton spinning other than in mills |
| 240 | Preparation of raw wool, silk, artificial/synthetic textile fibres for spinning |
| 25 | Manufacture of jute and other vegetable fibre textiles (except cotton and coir) |
| 270 | Sawing and planning of wood (other than plywood) |
| 271 | Manufacture of veneer sheets, plywood and their products |
| 272 | Manufacture of structural wooden goods (including treated timber) such as beams, posts, doors and windows (excluding hewing and rough shaping of poles, bolts and other wood material which is classified under logging) |
| 274 | Manufacture of wooden industrial goods n.e.c. |
| 275 | Manufacture of cork and cork products |
| 280 | Manufacture of pulp, paper and paper board including manufacture of newsprint |
| 281 | Manufacture of containers and boxes of paper or paper board |
| 290 | Tanning, curing, finishing, embossing and japanning of leather |
| 300 | Manufacture of industrial organic and inorganic chemicals |
| 301 | Manufacture of fertilizers and pesticides |
| 302 | Manufacture of plastics in primary forms; manufacture of synthetic rubber |
| 303 | Manufacture of paints, varnishes and related products; artists' colours and ink |
| 306 | Manufacture of man-made fibres |
| 308 | Manufacture of explosives; ammunition and fire-works |
| 309 | Manufacture of chemical products not elsewhere classified |
| 312 | Manufacture of rubber products not elsewhere classified |
| 313 | Manufacture of plastic products not elsewhere classified |
| 314 | Manufacture of refined petroleum products |
| 315 | Bottling of natural gas or liquified petroleum gas |
| 318 | Manufacture of coke oven products |
| 319 | Manufacture of other coal and coal tar products not elsewhere classified |
| 320 | Manufacture of refractory products and structural clay products |
| 324 | Manufacture of cement, lime and plaster |
| 325 | Manufacture of mica products |
| 326 | Stone dressing and crushing; manufacture of structural stone goods and stone ware |

| | |
|---|---|
| 327 | Manufacture of asbestos cement and other cement products |
| 329 | Manufacture of miscellaneous non-metallic mineral products not elsewhere classified |
| 33 | Basic metal and alloys industries |
| 344 | Forging, pressing, stamping and roll-forming of metal; powder metallurgy |
| 345 | Treatment or coating of metals; general mechanical engineering on a sub-contract basis |
| 361 | Manufacture of insulated wires and cables, including manufacture of optical fibre cables |

# CONSUMER NON-DURABLE GOODS INDUSTRY GROUP

| *NIC Code (1987)* | *Industries* |
|---|---|
| 20-1 | Manufacture of food products |
| 22 | Manufacture of beverages, tabacco and related products |
| 23 | Manufacture of cotton textiles except NIC 230 and 231 |
| 24 | Manufacture of wool, silk and man-made fibre textiles except NIC 240 |
| 26 | Manufacture of textile products (including wearing apparel) |
| 28 | Manufacture of paper and paper products and printing, publishing and allied industries except NIC 280 and 281 |
| 29 | Manufacture of leather and products of leather, fur and substitutes of leather except NIC 290 |
| 304 | Manufacture of drugs, medicines and allied products |
| 305 | Manufacture of perfumes, cosmetics, lotions, hair dressings, toothpastes, soap in any form, detergents, shampoos, shaving products, washing and cleaning preparations and other toilet preparations |
| 307 | Manufacture of matches |
| 311 | Manufacture of footwear made primarily of vulcanised or moulded rubber and plastics |
| 316 | Manufacture of refined petroleum products not elsewhere classified |
| 321 | Manufacture of glass and glass products |
| 322 | Manufacture of earthen and plaster products |
| 323 | Manufacture of non-structural ceramic ware |
| 362 | Manufacture of accumulators, primary cells and primary batteries |
| 363 | Manufacture of electric lamps |
| 368 | Manufacture of electronic valves and tubes and other electronic computers n.e.c. |
| 385 | Manufactures of sports and athletic goods |
| 387 | Manufacture of stationery articles n.e.c. |

# CONSUMER DURABLE GOODS INDUSTRY GROUP

| *NIC Code (1987)* | *Industries* |
|---|---|
| 273 | Manufacture of wooden and cane boxes, crates, drums, barrels and other containers, baskets and other wares made entirely or mainly of cane, rattan, reed, bamboo, willow, fibres, leaves and grass |
| 276 | Manufacture of wooden furniture and fixtures |
| 277 | Manufacture of bamboo and cane furniture and fixtures |
| 279 | Manufacture of products of wood, bamboo, cane, reed and grass n.e.c. |
| 310 | Tyre and tube industries |
| 342 | Manufacture of furniture and fixtures primarily of metals |
| 346 | Manufacture of metal cutlery, utensils and kitchenware |
| 349 | Manufacture of metal products (except machinery and equipment) not elsewhere classified |
| 355 | Manufacture of refrigerators, air-conditioners and firefighting equipment and their parts and accessories |
| 358 | Manufacture of office, computing and accounting machinery and parts |
| 359 | Manufacture of special purpose machinery/equipment, their components and accessories n.e.c. |
| 364 | Manufacture of electrical fans and electrical/electro-thermic domestic appliances and parts thereof |
| 365 | Manufacture of apparatus for radio broadcasting, television transmission radar apparatus and radio-remote control apparatus and apparatus for radio/line telephony and line telegraphy |
| 366 | Manufacture of television receivers; reception apparatus for radio broadcasting, radio telephony/telegraphy, video recording or reproducing apparatus, turn-tables, record players, cassette players and other sound reproducing apparatus, sound recording apparatus, microphones, loudspeakers, amplifiers and sound amplifiers and prerecorded audio/video records/tapes |
| 374 | Manufacture of motor cars and other motor vehicles principally designed for the transport of less than 10 persons |
| 375 | Manufacture of motorcycles and scooters and parts (including three-wheeler) |
| 376 | Manufacture of bicycles, cycle-rickshaws and parts |
| 381 | Manufacture of photographic, cinematographic and optical goods and equipment (excluding photochemicals, sensitised paper and film) |
| 382 | Manufacture of watches and clocks |
| 383 | Manufacture of jewellery and related articles |
| 386 | Manufacture of musical instruments |
| 389 | Manufacture of miscellaneous products not elsewhere classified |

# Contributors

R. Nagaraj is Associate Professor, Indira Gandhi Institute of Development Research.

Shuji Uchikawa is Visiting Faculty, Indira Gandhi Institute of Development Research, and Researcher, Institute of Developing Economies-JETRO.

Rajendra R. Vaidya is Associate Professor, Indira Gandhi Institute of Development Research.

Tirthankar Roy is Professor, Gokhale Institute of Politics and Economics.

K.V. Ramaswamy is Associate Professor, Indian Institute of Management, Indore.

# Index